Lecture Notes in Chemistry

Edited by G. Berthier M. J. S. Dewar H. Fischer
K. Fukui G. G. Hall H. Hartmann H. H. Jaffé J. Jortner
W. Kutzelnigg K. Ruedenberg E. Scrocco

28

G.S. Ezra

Symmetry Properties of Molecules

Springer-Verlag
Berlin Heidelberg New York 1982

Author

G.S. Ezra

Department of Chemistry, The University of Chicago

5735 South Ellis Avenue, Chicago, Illinois 60637, USA

ISBN-13: 978-3-540-11184-9 e-ISBN-13: 978-3-642-93197-0

DOI: 10.1007/978-3-642-93197-0

Softcover reprint of the hardcover 1st edition 1982

2152/3140-543210

Table of Contents

Acknowledgments

It is a pleasure to thank Dr. P.W. Atkins for his constant guidance and encouragement during the course of my work on molecular symmetry, and Professor R. S. Berry, both for sharing his stimulating insights into the phenomenon of non-rigidity and for generous support of the preparation of the final typescript. I have also had many fruitful conversations with Dr. G.A. Natanson, Dr. B.J. Howard, Professor H.P. Fritzer, Dr. P.R. Taylor and Dr. S.P. Keating. I am grateful to Dr. S.L. Altmann and particularly to Dr. B.T. Sutcliffe for their kind encouragment to make this review available to a wider audience. Finally, I must thank Fred Flowers for his very skillful preparation of the camera-ready typescript, and my wife Sally for her invaluable assistance with proof-reading.

The financial support of the U.K. Science Research Council and the Governing Body of Christ Church, Oxford, is gratefully acknowledged.

The preparation of this manuscript was supported in part by a Grant from the National Science Foundation.

Introduction

The aim of the present article is to give a critical exposition of the theory of the symmetry properties of rigid and nonrigid molecules. Despite the fact that several accounts of the subject, both technical and didactic, are now available, and despite the extensive discussion of nonrigid molecule symmetry that has been going on since the classic papers of Hougen and Longuet-Higgins, there remains a need for a unifying survey of the problem. Previous treatments have tended to emphasize one or the other particular viewpoint at the expense of a broader view.

Renewed interest in the details of the symmetry classification of rotation-vibration states of highly symmetric (octahedral) molecules has led to a reexamination of the relation between conventional point group operations and permutations of identical nuclei in rigid molecules, together with a clarification of the fundamental role of the Eckart constraints and associated Eckart frame. As is shown below, analogous insights can also be obtained in the case of nonrigid molecule symmetry, where the Eckart-Sayvetz conditions provide a natural generalization of the Eckart constraints.

The importance of particular definitions of the 'molecule-fixed' frame in the theory of molecular symmetry can be better appreciated by examining their dynamical origin. Chapter 1 is therefore devoted to a description of the derivation of the usual Wilson-Howard-Watson form of the molecular Hamiltonian, together with its generalization to nonrigid molecules. Particular attention is given to the introduction of molecular models and use of the Eckart and Eckart-Sayvetz constraints. Some other approaches to nonrigid molecule dynamics are also examined.

After a brief review of the fundamental symmetries of the complete molecular Hamiltonian, Chapter 2 gives a treatment of the symmetry properties of rigid molecules based upon the static molecular model, following closely that of Louck and Galbraith. Both the interpretation of feasible nuclear permutations and the invariance group of the Eckart frame are discussed in detail. The important problem of a correct definition of the parity of rotational wavefunctions, and hence of molecular systems, is also considered. The formal theory is illustrated by application to the symmetry classification of states of diatomic molecules.

Chapter 3 then develops a unified approach to the symmetry properties of nonrigid molecules. The formalism is based explicitly upon the properties of the semi-rigid molecular model, and is a straightforward generalization of the theory given for rigid molecules in Chapter 2. A symmetry group of the semi-rigid model is defined, and identified as the nonrigid molecule group. Induced transformations of Born-Oppenheimer variables result in feasible permutations of nuclei. It is shown that previous approaches can thereby be encompassed within a unified scheme.

Having dealt with fundamental matters concerning the nature of symmetry operations in nonrigid molecules, attention is turned to the related technical problem of handling the large symmetry groups involved. An investigation of the structure of nonrigid molecule symmetry groups is clearly important for the practical implementation of the theory. The formalism developed in Chapter 3 is well suited to the recognition and exploitation of nontrivial structure in nonrigid molecule groups. As recognized early on by McIntosh and by Altmann, these groups can often be written as semi-direct products. The systematic theory of semi-direct products is briefly reviewed in an appendix, while Chapter 4 applies the theory to derive character tables for various nonrigid molecule groups. Recognition of semi-direct product structure allows a straightforward correlation to be made between the irreducible representations of rigid and nonrigid molecule symmetry groups. Such correlations can be exploited in the general theory of vibrations in nonrigid molecules, but the details of this application have yet to be worked out.

The appendices develop in detail several important topics touched upon in the text. A glossary of mathematical symbols and abbreviations has been included for the convenience of the reader.

Chapter 1. The Molecular Hamiltonian

We begin by considering the formulation of Hamiltonians for molecules both rigid and nonrigid. Although we are primarily concerned with molecular symmetry groups and their structure, an excursion into the theory of molecular dynamics is necessary for several reasons. First of all, while it is true that notions of molecular symmetry ultimately derive from the essential indistinguishability of identical micro-particles (nuclei), which is a non-dynamical concept, in practice our point of view is of necessity dynamics dependent. Thus, the <u>feasibility</u> of a particular transformation [1] - in other words, the extent to which a given symmetry is manifest in a given experiment - is obviously entirely contingent upon the forces acting within the molecule, for a finite experimental resolution/observation time. Conversely, since it is not possible at present to set up and solve nontrivial many-particle (nuclear or molecular) problems using a set of coordinates displaying <u>all</u> permutational symmetries in a simple fashion [2,3], the very way in which we approach the dynamics is directly determined by intuitive ideas concerning feasibility. One of the points we shall seek to emphasize throughout our work is the close relation between descriptions of symmetry and dynamics. Again, a knowledge of the transformation from cartesian to molecular (Born-Oppenheimer) coordinates used to rewrite the Hamiltonian is essential when we come to the important practical problem of determining the induced action of permutations of identical nuclei upon molecular wavefunctions.

The general problem of the derivation of the quantum-mechanical molecular Hamiltonian operator expressed in Born-Oppenheimer coordinates has recently been reviewed by Makushkin and Ulenikov [4] and we will not trace the history of this subject here. For the details of the procedure, we have chosen to follow the recent account by Sørensen [5], which has, in our view, several attractive features. Thus, Sørensen's work emphasizes the fact that it is possible to develop a treatment of nonrigid molecule (NRM) dynamics entirely analogous to that for quasi-rigid species, by introducing the concept of the semi-rigid molecular model (§1.4). In addition, the stress laid upon the Eckart and Eckart-Sayvetz conditions reflects the importance we shall ascribe to them in our discussions of molecular symmetry in Chapters 2 and 3. Hence, a careful discussion of the molecular Hamiltonian from this viewpoint is useful preparation for our later work, in which we shall draw on many of the ideas introduced here.

In this chapter, we are therefore concerned with the passage from the classical expression for the molecular energy in lab-fixed cartesian coordinates (§1.1) to the Wilson-Howard-Watson form of the quantum-mechanical Hamiltonian (§1.3) and its natural generalization to nonrigid systems (§1.4). A brief discussion of some other approaches to the dynamics of highly nonrigid systems is also given (§1.5).

1.1 The Molecular Kinetic Energy

Classically, the total non-relativistic energy of an isolated molecule consisting of N nuclei and N_ε electrons can be written

$$E_{mol} = \frac{1}{2}\sum_{\alpha} m_\alpha^{-1}\dot{R}_i^\alpha \dot{R}_i^\alpha + (\frac{1}{2}m)\sum_{\varepsilon} \dot{R}_i^\varepsilon \dot{R}_i^\varepsilon + V_{NN}(\underset{\sim}{R}^\alpha) + V_{NE}(\underset{\sim}{R}^\alpha;\underset{\sim}{R}^\varepsilon) + V_{EE}(\underset{\sim}{R}^\varepsilon) \qquad 1.1$$

where we introduce the following notation: $\{\hat{\underset{\sim}{\ell}}_i;\ i = x,y,z\}$ is the laboratory coordinate frame, right-handed by definition; $\underset{\sim}{R}^\alpha \equiv$ position of nucleus α, charge z_α, mass m_α, $1 \leq \alpha \leq N$, $R_i^\alpha \equiv \hat{\underset{\sim}{\ell}}_i\cdot\underset{\sim}{R}^\alpha$, $\dot{R}_i^\alpha \equiv d/dt_{lab}R_i^\alpha$; $\underset{\sim}{R}^\varepsilon \equiv$ position of electron ε, mass m, $1 \leq \varepsilon \leq N_\varepsilon$, $R_i^\varepsilon \equiv \hat{\underset{\sim}{\ell}}_i\cdot\underset{\sim}{R}^\varepsilon$, $\dot{R}_i^\varepsilon \equiv d/dt_{lab}R_i^\varepsilon$; V_{NN}, V_{NE} and V_{EE} are the nucleus/nucleus, nucleus/electron and electron/electron coulomb potential energies, respectively and there is a summation convention for repeated indices i,j etc.

Our task now is to adopt some suitable set of 'molecule-fixed' coordinates, to express the molecular energy in terms of these new coordinates, to obtain a Hamiltonian form for the energy, and finally to quantize the classical Hamiltonian using a suitable quantization rule (this is the traditional route: cf. [6], Chapter 11; note, however, that a more direct approach is taken in [4,7]).

The starting point is the coordinate transformation shown in Figure 1.1 [5,8]

$$R_i^\alpha = R_i + C_{ij}r_j^\alpha(q_\lambda) \qquad 1.2a$$

$$R_i^\varepsilon = R_i + C_{ij}r_j^\varepsilon \qquad 1.2b$$

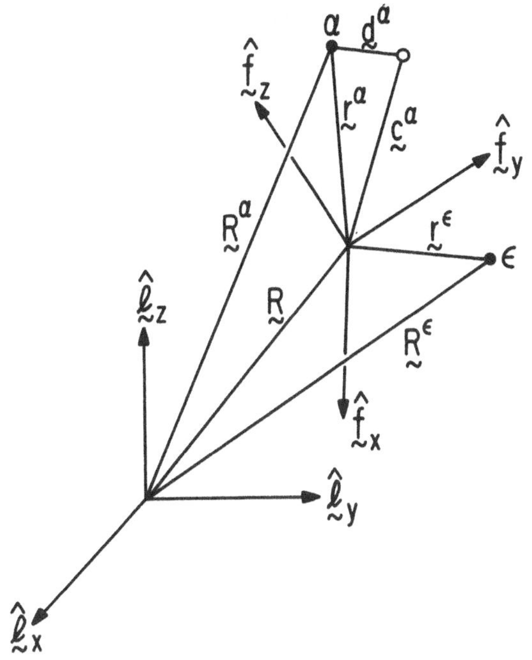

Figure 1.1 The transformation from lab-fixed to molecular (Born-Oppenheimer) coordinates

where $\underset{\sim}{R} \equiv (\sum_{\alpha} m_{\alpha} \underset{\sim}{R}^{\alpha} + m \sum_{\varepsilon} \underset{\sim}{R}^{\varepsilon})/(\sum_{\alpha} m_{\alpha} + mN_{\varepsilon})$ is the molecular centre of mass, $R_i \equiv \hat{\underset{\sim}{\ell}}_i \cdot \underset{\sim}{R}$, $\{q_{\lambda};\ \lambda = 1...3N\text{-}6\}$ are 3N-6 independent internal coordinates, $\{\hat{\underset{\sim}{f}}_i\}$ is the 'molecule-fixed' orthonormal coordinate frame, and $C_{ij} \equiv \hat{\underset{\sim}{\ell}}_i \cdot \hat{\underset{\sim}{f}}_j$ are elements of the direction-cosine matrix C.

Equation 1.1 is a transformation of the electronic and nuclear coordinates into a frame $\{\hat{\underset{\sim}{f}}_i\}$ with origin at the molecular centre of mass, 'rotating with the molecule' in some sense; the 3N components $\{r_j^{\alpha}\}$ depend on the 3N-6 independent generalized coordinates $\{q_{\lambda}\}$. The specification of a non-inertial frame carrying the total orbital angular momentum of the system is a major problem in the general theory of collective motion in many-particle systems (cf. [2] and references therein). In the molecular case, however, it is often appropriate to define the orientation of $\{\hat{\underset{\sim}{f}}_i\}$ using the Eckart [9] or Eckart-Sayvetz [10] conditions, so that we can refer to $\{\hat{\underset{\sim}{f}}_i\}$ as the <u>Eckart frame</u> (§1.2).

Although the laboratory frame $\{\hat{\underset{\sim}{\ell}}_i\}$ is by definition right-handed, the direction-cosine matrix C is allowed to have determinant +1 <u>or</u> -1, so that the sense of the molecule-fixed frame can be opposite to that of the laboratory frame, i.e., left-handed. The matrix C is therefore an element of the orthogonal matrix group O(3)

$$C \in O(3) : C_{ij}C_{ij'} = \delta_{jj'} , \ \det C = \pm 1. \qquad 1.3$$

We set

$$C \equiv (\det C)C' \qquad 1.4a$$

where C' is a proper rotation matrix

$$C' \in SO(3) , \ \det C' = +1 \qquad 1.4b$$

and is a function of 3 independent parameters such as the Euler angles (Appendix 1).

The only consequence of C being an element of O(3) for the derivation of the Hamiltonian is that some care is necessary when projecting the components of axial vectors, such as vector products, from the laboratory frame onto $\{\hat{\underset{\sim}{f}}_i\}$ (as pointed out by Husson [12]). The resulting factors of (det C) are however omitted in the rest of this chapter for the sake of clarity.

The extra freedom in the rotational coordinates is motivated by several considerations. First, as discussed in §1.2, the phenomenon of 'Eckart frame inversion' [13] shows that when large-amplitude nuclear motions are taken into account the Eckart frame does not necessarily have the same handedness as the laboratory frame. Also, Louck [7] has shown that the 'internal' and 'external' angular momenta appearing in the molecular problem can be expressed directly in terms of the direction-cosine matrix C. Not only are these expressions independent of any specific parametrization of C', but they are also invariant, as is the corresponding Hamiltonian, under the inversion $\mathcal{I}: C \to -C$ (cf. Appendix 2). As will be seen in Chapter 2, the problem of determining the transformation properties of rotational wavefunctions under improper rotations of axes is important for a detailed understanding of molecular symmetry. Using simple vector-coupling arguments (given in Appendix 2), it is possible to write rotational wavefunctions as homogeneous polynomials in the elements of the matrix C. There is then no ambiguity concerning either the parity of rotational wavefunctions, and hence of molecular systems, or the induced action of improper rotations.

Returning to the coordinate transformation 1.2, it should be emphasized that at this point the 3N-6 independent internal coordinates $\{q_\lambda\}$ are entirely general in nature, and do not necessarily describe displacements of nuclei from notional 'equilibrium positions'. In fact, it is only required that the transformation 1.2 be invertible, and that variation of the $\{q_\lambda\}$ should leave the nuclear centre of mass invariant [8].

In the work to follow we identify the molecular centre of mass $\underset{\sim}{R}$ with the nuclear centre of mass $\underset{\sim}{R}^{nuc}$

$$\underset{\sim}{R} \sim \underset{\sim}{R}^{nuc} \equiv \left(\sum m_\alpha \underset{\sim}{R}^\alpha\right)/\sum m_\alpha . \qquad 1.5$$

This is a conventional approximation [18], which has not however been made by Howard and Moss [19], who have explicitly derived the small (mass-polarization or recoil) correction terms arising from the fact that $\underset{\sim}{R} \neq \underset{\sim}{R}^{nuc}$. We also ignore the motions of the electrons completely. It is therefore assumed that the nuclei move in a translation/rotation invariant potential field $V(q_\lambda)$, corresponding to the 'potential energy surface' associated with a particular electronic state. This is the viewpoint of the conventional Born-Oppenheimer approximation [20, 21], and means that the Hamiltonian we shall derive should be regarded as an effective molecular Hamiltonian corresponding to a given electronic state [22,23].

We shall not be particularly concerned with the details of quantizing the molecular Hamiltonian [7, 24-26] or of obtaining hermitian operators [27], and so work mainly within the framework of classical mechanics.

From equation 1.2, the velocity of nucleus α is

$$d/dt_{lab}\, \underset{\sim}{R}^{\alpha} = d/dt_{lab}\, \underset{\sim}{R} + \underset{\sim}{\omega} \wedge \underset{\sim}{r}^{\alpha} + \sum_{\lambda} \frac{\partial \underset{\sim}{r}^{\alpha}}{\partial q_{\lambda}} \dot{q}_{\lambda} \qquad 1.6$$

where the angular velocity vector $\underset{\sim}{\omega}$ is defined by

$$d/dt_{lab}\, \hat{\underset{\sim}{f}}_i \equiv \underset{\sim}{\omega} \wedge \hat{\underset{\sim}{f}}_i \qquad 1.7$$

so that

$$\underset{\sim}{\omega} \wedge \underset{\sim}{r}^{\alpha} = \underset{\sim}{\omega} \wedge \hat{\underset{\sim}{f}}_j r_j^{\alpha} \; . \qquad 1.8$$

The sense of the vector product is determined in the lab frame, and

$$\underset{\sim}{\omega} \wedge \underset{\sim}{r}^{\alpha} = (\hat{\underset{\sim}{f}}_j \wedge \hat{\underset{\sim}{f}}_k) \omega_j r_k^{\alpha} = (\det C) \hat{\underset{\sim}{f}}_i \varepsilon_{ijk} \omega_j r_k^{\alpha} \qquad 1.9$$

with

$$\omega_j \equiv \hat{\underset{\sim}{f}}_j \cdot \underset{\sim}{\omega} \; . \qquad 1.10$$

Defining the 3N-dimensional velocity vector

$$(v_1, v_2, \ldots, v_{3N}) \equiv (\dot{R}_x, \dot{R}_y, \dot{R}_z, \omega_x, \omega_y, \omega_z, q_1, \ldots, q_{3N-6}) \qquad 1.11$$

we can write

$$\dot{\underset{\sim}{R}}^{\alpha} \equiv \sum_{\nu=1}^{3N} \underset{\sim}{t}_{\alpha,\nu} v_{\nu} = \sum_{\nu} \hat{\underset{\sim}{\ell}}_i t_{\alpha i,\nu} v_{\nu} \qquad 1.12$$

where the transformation coefficients $[t_{\alpha i,\nu}]$ form a 3N by 3N matrix **t**. In detail, from 1.6 the t-vectors are

Translation	$\underset{\sim}{t}_{\alpha,Ti} = \hat{\underset{\sim}{\ell}}_i$	$i = x,y,z$	1.13a
Rotation	$\underset{\sim}{t}_{\alpha,Ri} = (\hat{\underset{\sim}{f}}_i \wedge \underset{\sim}{r}^{\alpha})$	$i = x,y,z$	1.13b
Internal	$\underset{\sim}{t}_{\alpha,\lambda} = \partial \underset{\sim}{r}^{\alpha} / \partial q_{\lambda}$.		1.13c

The nuclear kinetic energy is then

$$2T = \sum_{\alpha} m_{\alpha} \dot{R}_i^{\alpha} \dot{R}_i^{\alpha} = \sum_{\nu,\nu'} K_{\nu\nu'} v_{\nu} v_{\nu'} \qquad 1.14$$

with the matrix

$$K_{\nu\nu'} \equiv \sum m_{\alpha} \underset{\sim}{t}_{\alpha,\nu} \cdot \underset{\sim}{t}_{\alpha,\nu'} \qquad 1.15$$

i.e.,

$$\mathbf{K} = \tilde{\mathbf{t}}\,\mathbf{m}\,\mathbf{t} \qquad \mathbf{m} = \text{mass matrix.} \qquad 1.16$$

In the assumed absence of nuclear velocity dependent potentials, we introduce generalized momenta and 'quasi-momenta' in the usual fashion (recall that the components of the angular velocity vector are not conjugate to any coordinates - the relation between the angular velocities and the time derivatives of, for example, the Euler angles, is non-integrable [29])

$$P_{\nu} \equiv \partial T/\partial v_{\nu} = \sum_{\nu'} K_{\nu\nu'} v_{\nu'} \qquad 1.17$$

and obtain

$$2T = \sum_{\nu} P_{\nu} v_{\nu} \qquad 1.18$$

which is the form-invariant expression for the kinetic energy. Defining the inverse of **K**

$$\mathbf{G} \equiv \mathbf{K}^{-1} \qquad 1.19$$

we have

$$v_{\nu} = \sum_{\nu'} G_{\nu\nu'} P_{\nu'} \qquad 1.20$$

which yields the Hamiltonian form

$$2T = \sum G_{\nu\nu'} P_{\nu} P_{\nu'} \ . \qquad 1.21$$

At this point it is possible to 'quantize' the classical molecular Hamiltonian. In essence, this involves using the formula for the 3N-dimensional Laplace-Beltrami operator in generalized coordinates (corresponding to the nuclear kinetic energy) expressed in terms of the elements of the matrix **G** appearing in 1.21 - this is known as the Podolsky [24] quantization procedure. Careful attention has to be given to problems concerning volume elements for integration of wavefunctions, and the presence of quasi-momenta in 1.21 [27].

To sum up, the kinetic energy 1.21 is obtained in 3 steps [5]:

a) Form the elements of the matrix [**t**] as functions of the generalized coordinates $\{q_\lambda\}$ (1.13).

b) Multiply matrices to obtain **K** (1.16).

c) Invert **K** to obtain the important matrix **G** (1.19).

It is the last operation of inversion that is difficult to perform explicitly, since **K** is, in general, q-dependent (although see [44]).

In the light of this, Sørensen has suggested the following procedure involving <u>momentum transformation</u>:

a') Obtain the expression for the nuclear momentum

$$\underset{\sim}{p}^\alpha \equiv \partial T/\partial \dot{\underset{\sim}{R}}^\alpha \equiv \sum_\nu P_\nu \underset{\sim}{s}_{\nu,\alpha} \equiv \sum_\nu P_\nu s_{\nu,\alpha j} \hat{\underset{\sim}{\ell}}_j \qquad 1.22$$

b') Write the nuclear kinetic energy as

$$2T = \sum_\alpha m_\alpha^{-1} \underset{\sim}{p}^\alpha \cdot \underset{\sim}{p}^\alpha = \sum_{\nu\nu'} G_{\nu\nu'} P_\nu P_{\nu'} \qquad 1.23$$

where the G-matrix is now given <u>directly</u> in terms of the s-vectors as

$$G_{\nu\nu'} = \sum m_\alpha^{-1} \underset{\sim}{s}_{\nu,a} \cdot \underset{\sim}{s}_{\lambda',\alpha} \,. \qquad 1.24$$

As the notation suggests, this is a direct generalization of the method used to obtain the <u>vibrational</u> kinetic energy in the Wilson FG-formalism [6]; however, here it is possible to deal with all nuclear degrees of freedom, not just the vibrational, together on the same footing, in the absence of any restriction to 'small-amplitude' motions.

Noting that

$$2T = \sum_\alpha \underset{\sim}{p}^\alpha \cdot \dot{\underset{\sim}{R}}^\alpha = \sum_{\alpha\nu,\nu'} P_\nu \underset{\sim}{s}_{\nu,\alpha} \cdot \underset{\sim}{t}_{\alpha,\nu'} v_{\nu'} \qquad 1.25$$

we must have from 1.18

$$\sum_\alpha \underset{\sim}{s}_{\nu,\alpha} \cdot \underset{\sim}{t}_{\alpha,\nu'} = \delta_{\nu\nu'} \qquad 1.26$$

i.e., the coefficients $[s_{\nu,\alpha j}]$ form a 3N by 3N matrix which is the left-inverse of **t**. The orthogonality relations 1.26 are very important in the development of the Hamiltonian. Thus, from the expressions for the translational and rotational t-vectors we note that

$$\sum_\alpha \underset{\sim}{s}_{\lambda,\alpha} \cdot \underset{\sim}{t}_{\alpha,Ti} = 0 \qquad \text{for all } \lambda, i \qquad 1.27a$$

implies

$$\sum_{\alpha} \underset{\sim}{s}_{\lambda,\alpha} = \underset{\sim}{0} \qquad \text{for all } \lambda \qquad 1.27b$$

and that

$$\sum_{\alpha} \underset{\sim}{s}_{\lambda,\alpha} \cdot \underset{\sim}{t}_{\alpha,Ri} = 0 \qquad \text{for all } \lambda,i \qquad 1.28a$$

implies

$$\sum_{\alpha} \underset{\sim}{r}^{\alpha} \wedge \underset{\sim}{s}_{\lambda,\alpha} = \underset{\sim}{0} \qquad \text{for all } \lambda \qquad 1.28b$$

The relations 1.27b and 1.28b are generalized Malhiot-Ferigle conditions [5,8,29], and show that the internal coordinates $\{q_\lambda\}$ are invariant under translations and rotations of the nuclear configuration $\{\underset{\sim}{R}^{\alpha}\}$, as required [8,30].

The outstanding problem is the evaluation of the s-vectors. From the relations (cf. 1.12)

$$\dot{R}_i = \sum_{\alpha} \underset{\sim}{s}_{Ti,\alpha} \cdot \dot{\underset{\sim}{R}}^{\alpha} \qquad 1.29a$$

$$\dot{q}_\lambda = \sum \underset{\sim}{s}_{\lambda,\alpha} \cdot \dot{\underset{\sim}{R}}^{\alpha} \qquad 1.29b$$

it follows immediately that

Translation $$\underset{\sim}{s}_{Ti,\alpha} = (m_\alpha/M)\, \hat{\underset{\sim}{\ell}}_i \qquad 1.30a$$

Internal $$\underset{\sim}{s}_{\lambda,\alpha} = \underset{\sim}{\nabla}^{\alpha} q_\lambda \equiv \hat{\underset{\sim}{\ell}}_j\, \partial q_\lambda/\partial R^{\alpha}_j \qquad 1.30b$$

where, for a geometrically-defined internal coordinate q_λ we can evaluate the derivative $(\partial q_\lambda/\partial R^{\alpha}_j)$ as a function of the instantaneous nuclear configuration $[R^{\alpha}_j]$. In §1.2, a more convenient expression for $\underset{\sim}{s}_{\lambda,\alpha}$ is obtained through use of rectilinear coordinates for rigid molecules.

How, then, to evaluate the rotational s-vectors $\underset{\sim}{s}_{Ri,\alpha}$? Consideration of the orthogonality condition

$$\sum_{\alpha} \underset{\sim}{s}_{Ti,\alpha} \cdot \underset{\sim}{t}_{\alpha,\lambda} = 0 \qquad \text{for all } \lambda,i \qquad 1.31$$

together with 1.13c and 1.30a shows that

$$\sum_{\alpha} m_\alpha(\partial \underset{\sim}{r}^{\alpha}/\partial q_\lambda) = \underset{\sim}{0} \qquad \text{for all } \lambda. \qquad 1.32$$

However, as recognized by Sørensen [5], this is just a differential consequence of the centre of mass condition (translational constraint)

$$\sum_{\alpha} m_\alpha \underset{\sim}{r}^{\alpha} = \underset{\sim}{0}\,. \qquad 1.33$$

This suggests that the orthogonality conditions

$$\sum_{\alpha} \underset{\sim}{s}_{Ri,\alpha} \cdot \underset{\sim}{t}_{\alpha,\lambda} = 0 \qquad \text{for all } \lambda, i \qquad 1.34$$

are equivalent to the differential consequences of the rotational constraints

$$C^{(i)}(r_j^{\alpha}) = 0 \qquad i = x,y,z \qquad 1.35$$

which serve to define the orientation of the molecule-fixed frame. Thus, assuming the $C^{(i)}$ to be differentiable functions of the nuclear coordinates, we have

$$\partial C^{(i)}/\partial q_{\lambda} = \sum_{\alpha} (\partial C^{(i)}/\partial r_j^{a})(\partial r_j^{\alpha}/\partial q_{\lambda}) = \sum_{\alpha} \underset{\sim}{c}_{i,\alpha} \cdot \underset{\sim}{t}_{\alpha,\lambda} = 0 \qquad i = x,y,z \qquad 1.36$$

where we have defined the <u>constraint vectors</u>

$$\underset{\sim}{c}_{i,\alpha} \equiv \sum_{\alpha} \hat{\underset{\sim}{f}}_j (\partial C^{(i)}/\partial r_j^{\alpha}) \ . \qquad 1.37$$

Sørensen's results can now be stated (and are easily verified): the rotational s-vectors are given as

$$\underset{\sim}{s}_{Ri,\alpha} = \eta_{ii'} \underset{\sim}{c}_{i',\alpha} \qquad 1.38$$

where η is a 3 by 3 matrix <u>common to all nuclei in the molecule</u>, and is the inverse of the matrix **A**, where

$$\eta = \mathbf{A}^{-1} \ , \qquad A_{ii'} \equiv \sum_{\alpha} \underset{\sim}{c}_{i,\alpha} \cdot \underset{\sim}{t}_{\alpha,Ri'} \ . \qquad 1.39$$

This result is very important, since it shows that, to derive the G-matrix directly, we need only invert a single 3 by 3 matrix **A**. The rotational part of the G-matrix is therefore

$$G_{Ri,Ri'} = \eta_{ij} I_{jj'} \tilde{\eta}_{j'i'} \qquad 1.40a$$

where

$$I_{jj'} \equiv \sum m_{\alpha}^{-1} \underset{\sim}{c}_{j,\alpha} \cdot \underset{\sim}{c}_{j',\alpha} \qquad 1.40b$$

i.e.,

$$[\mathbf{G}]_{rot} = \eta \ I \ \tilde{\eta} \qquad 1.40b$$

so that we have obtained in a very general fashion the factorization of $[\mathbf{G}]_{rot}$ (otherwise known as the μ-tensor) previously noted [26] for the special case of rigid molecules. In the next section, I is identified as the inertia tensor of the nuclear reference configuration.

1.2 Quasi-Rigid Molecules

In this section, the general coordinate transformation 1.2 is specialized to a form appropriate for the description of the dynamics of quasi-rigid (otherwise known as 'rigid') molecules. It is thereby possible to derive, in §1.3, the familiar Wilson-Howard-Watson Hamiltonian for non-linear rigid molecules. Important features of the specialization to rigid species are (cf. [5]): 1) Introduction of the static molecular model [17], hence the notion of molecular structure [3,31-33]. 2) Use of Eckart's rotational constraints to define the rotational coordinates C_{ij} [9,17]. 3) Use of rectilinear vibrational coordinates to define the displacements of nuclei from their equilibrium positions. We now consider each in turn:

1) The intuitive dynamical picture underlying the conventional treatment of rigid molecule dynamics is very familiar [6]. The nuclei are assumed to execute rapid, small-amplitude vibrations about a fixed, 'equilibrium' configuration that is itself undergoing overall rotation. With this in mind, we write

$$\underset{\sim}{r}^{\alpha} \equiv (\underset{\sim}{R}^{\alpha} - \underset{\sim}{R}) = \underset{\sim}{c}^{\alpha} + \underset{\sim}{d}^{\alpha} \qquad 1.41$$

where $\underset{\sim}{c}^{\alpha}$ is to be interpreted as the equilibrium position of nucleus α, and $\underset{\sim}{d}^{\alpha}$ is the displacement from equilibrium. Recall that it is necessary somehow to define a molecule-fixed frame $\{\hat{\underset{\sim}{f}}_i\}$; in this coordinate frame, the vectors $\underset{\sim}{c}^{\alpha}$ are constant, by definition

$$\underset{\sim}{c}^{\alpha} \equiv \hat{\underset{\sim}{f}}_i a_i^{\alpha} \qquad 1.42$$

so that the coordinate transformation 1.2 can be written (Figure 1.1)

$$R_i^{\alpha} - R_i = C_{ij}(a_j^{\alpha} + d_j^{\alpha}) \qquad 1.43a$$

$$d_j^{\alpha} \equiv \hat{\underset{\sim}{f}}_j \cdot \underset{\sim}{d}^{\alpha} . \qquad 1.43b$$

The 3N constants $\{a_j^{\alpha}\}$ define the static molecular model $\mathcal{A}$, which is the set of triples

$$\mathcal{A} \equiv \{(\underset{\sim}{a}^{\alpha}, z_{\alpha}, m_{\alpha}); \quad \alpha = 1,\ldots,N\} \qquad 1.44$$

where the vector $\underset{\sim}{a}^{\alpha}$ is associated with a nucleus of charge z_{α} and mass m_{α} (Louck and

Galbraith [17] introduced the notational distinction between the vector $\underset{\sim}{c}^{\alpha} \equiv \hat{\underset{\sim}{f}}_i a_i^{\alpha}$ and the static model vector $\underset{\sim}{a}^{\alpha}$; this indicates that, for any α, the three numbers a_j^{α} define an abstract vector $\underset{\sim}{a}^{\alpha}$, whereas $\underset{\sim}{c}^{\alpha}$ is a possible nuclear position vector in 'real' space spanned by the basis $\{\hat{\underset{\sim}{\ell}}_i\}$).

It is clear that the components $\{a_j^{\alpha}\}$ embody the concept of molecular structure, which makes its way into the theory <u>via</u> the coordinate transformation 1.43. A question immediately arising is: how do we determine relevant $\{a_j^{\alpha}\}$?

In principle, possible reference configurations for a molecule are determined by the following procedure: Consider the electronic Hamiltonian

$$\hat{H}_E \equiv \hat{T}_E + V_{NN} + V_{NE} + V_{EE} \qquad 1.45$$

depending only upon <u>fixed</u> nuclear positions $\{\underset{\sim}{X}^{\alpha}\}$, and not upon nuclear momenta. The eigenvalue equation

$$\hat{H}_E \chi_{e\ell}^{\nu}(\underset{\sim}{R}^{\varepsilon};\underset{\sim}{X}^{\alpha}) = W_{\nu}(\underset{\sim}{X}^{\alpha})\chi_{e\ell}^{\nu}(\underset{\sim}{R}^{\varepsilon};\underset{\sim}{X}^{\alpha}) \qquad 1.46$$

must then be solved to give the potential energy surfaces corresponding to the various electronic states labeled by the quantum number ν; that is, the electronic energies W_{ν} as functions of the nuclear configuration $\{\underset{\sim}{X}^{\alpha}\}$.

Choosing a particular quantum number ν (which labels the electronic state for which the vibration-rotation Hamiltonian to be derived is an effective Hamiltonian), we search for a minimum in the surface, denoted by the configuration $\{\underset{\sim}{X}_0^{\alpha}\}$. For a given molecule, there will in general be several local minima, corresponding to different isomers and so on [34, 119]. However, if there is a minimum at $\{\underset{\sim}{X}_0^{\alpha}\}$, then there is certainly a minimum at the symmetry-related configuration $\{\underset{\sim}{X}_0^{\bar{\alpha}}\}$, where the set of labels $\{\bar{\alpha}\}$ differs from the set $\{\alpha\}$ by a permutation of identical nuclei [35] (cf. Chapter 2).

A given configuration $\{\underset{\sim}{X}_0^{\alpha}\}$ defines an equilibrium structure <u>modulo</u> overall translation/rotation. However, by eliminating the nuclear centre of mass and adopting a principal-axis coordinate frame $\{\hat{\underset{\sim}{e}}_i\}$, so that

$$\underset{\sim}{X}_0^{\alpha} \equiv \underset{\sim}{a}^{\alpha} \equiv \hat{\underset{\sim}{e}}_i a_i^{\alpha} \qquad 1.47a$$

with

$$\sum m_{\alpha} a_i^{\alpha} = 0 \qquad i = x,y,z \qquad 1.47b$$

and $$\sum m_\alpha a_i^\alpha a_{i'}^\alpha = 0 \quad i \neq i' \quad , \qquad 1.47c$$

we obtain a particular static molecular model $\{a_i^\alpha\}$. This set of numbers is appropriate for substitution into 1.43.

The above account summarizes conventional wisdom on the subject; however, it has been pointed out that the $\{a_i^\alpha\}$ in 1.43 can be taken to be essentially arbitrary, unrelated to any notion of 'equilibrium geometry' [3,23]. Nevertheless, it should be recalled that the original work of Born and Oppenheimer [20] showed that a necessary condition for the validity of the usual perturbative scheme for separation of electronic, vibrational and rotational energies in the absence of electronic degeneracies is that the static molecular model defines an extremum (usually taken to be a minimum, since we require a locally stable structure), so that there seem to be significant limitations on the practical utility of alternative definitions of the $\{a_i^\alpha\}$ differing greatly from equilibrium configurations.

2) For any arbitrary many-particle system, the introduction of a non-inertial rotating frame leads to the appearance of Coriolis-type interactions between internal motions and overall rotations of the system [36]. A useful definition of the molecule-fixed frame should minimize such vibration/rotation coupling. However, it must be recognized that it is not possible to eliminate Coriolis interactions completely [9], since attempts to do so result in non-integrable (non-holonomic) constraints, as has been discovered in both molecular [8] and nuclear [37] physics (cf. however [38]). The Eckart conditions [5,9,17], on the other hand, eliminate Coriolis coupling only to first order in the nuclear displacements (§1.3), but lead to constraints that are easy to handle. In addition, the Eckart frame is of special significance in the theory of rigid molecule symmetry, as discussed in Chapter 2.

Aside from the centre of mass condition

$$\sum_\alpha m_\alpha \underset{\sim}{r}^\alpha = \sum_\alpha m_\alpha \underset{\sim}{d}^\alpha = \underset{\sim}{0} \qquad 1.48$$

the rotational Eckart conditions are

$$\sum_\alpha m_\alpha \underset{\sim}{c}^\alpha \wedge \underset{\sim}{r}^\alpha = \sum m_\alpha \underset{\sim}{c}^\alpha \wedge \underset{\sim}{d}^\alpha = \underset{\sim}{0} \; . \qquad 1.49$$

The 3 rotational constraints 1.35 are therefore

$$C^{(i)} = \varepsilon_{ijk} \sum_\alpha m_\alpha a_j^\alpha r_k^\alpha = 0 \qquad 1.50$$

and the constraint vectors become

$$\underset{\sim}{c}_{i,\alpha} = \varepsilon_{ijk} m_\alpha a_j^\alpha \hat{\underset{\sim}{f}}_k = \underset{\sim}{f}_i \wedge \underset{\sim}{c}^\alpha \qquad 1.51$$

and are <u>constant</u> in the frame $\{\hat{\underset{\sim}{f}}_i\}$.

The tensor I appearing in the rotational part of the G-matrix 1.40 is

$$I_{ii'} \equiv \sum_\alpha m_\alpha^{-1} \underset{\sim}{c}_{i,\alpha} \cdot \underset{\sim}{c}_{i',k}$$

$$= \varepsilon_{ijk}\varepsilon_{i'j'k} \sum m_\alpha a_j^\alpha a_{j'}^\alpha = I_{ii'}^0 \qquad 1.52$$

which is the inertia tensor of the nuclear reference configuration. Also, from 1.13b the elements of **A** are

$$A_{ii'} = \varepsilon_{ijk}\varepsilon_{i'j'k} \sum m_\alpha a_j^\alpha r_{j'}^\alpha , \qquad 1.53$$

where **A** is a symmetric matrix (Eckart conditions)

$$\mathbf{A} = \tilde{\mathbf{A}} , \ \boldsymbol{\eta} = \tilde{\boldsymbol{\eta}} \qquad 1.54$$

which is also linear in the nuclear displacements $\{d_j^\alpha\}$, so that

$$\mathbf{A} = \mathrm{I} = \mathrm{I}^0 \quad \text{at equilibrium.} \qquad 1.55$$

The relations 1.49,50 only define the Eckart frame $\{\hat{\underset{\sim}{f}}_i\}$ implicitly. However, in his original account ([9], cf. also [17,39]), Eckart gave a method for the explicit construction of the $\{\hat{\underset{\sim}{f}}_i\}$ in terms of the instantaneous configuration $\{\underset{\sim}{r}^\alpha\}$. Indeed, it is a central feature of the method that, although the constraints have a dynamical significance in that Coriolis coupling is minimized to first order, the orientation of the Eckart frame itself depends upon nuclear <u>positions</u> only (together with the static model).

Thus, defining the three <u>Eckart vectors</u> $\underset{\sim}{F}_i$

$$\underset{\sim}{F}_i \equiv \sum m_\alpha \underset{\sim}{r} a_i^\alpha \qquad i = x,y,z \qquad 1.56$$

the Eckart frame $\{\hat{\underset{\sim}{f}}_i\}$ is obtained from the $\underset{\sim}{F}_i$ by a process of symmetric orthonormalization [39]. Forming the Gram matrix $\boldsymbol{\Gamma}$

$$\Gamma_{ii'} \equiv \underset{\sim}{F}_i \cdot \underset{\sim}{F}_{i'} \qquad 1.57$$

we have

$$\hat{\underset{\sim}{f}}_i = \underset{\sim}{F}_{i'}(\Gamma^{-1/2})_{i'i} \qquad 1.58$$

(for convenience we consider only non-planar molecules; for planar static models there are only two linearly independent Eckart vectors, whereas for linear molecules there is only one [17]). It is simple to verify that the Eckart conditions hold for the frame vectors defined in 1.58, i.e.,

$$\underset{\sim}{F}_i \cdot \hat{\underset{\sim}{f}}_j = \underset{\sim}{F}_j \cdot \hat{\underset{\sim}{f}}_i \; . \qquad 1.58'$$

To calculate the square root of the positive definite Gram matrix Γ, it must first be diagonalized. If

$$\Lambda = U^{+}\Gamma U \quad , \quad \Gamma = U\Lambda U^{+} \qquad 1.59a$$

with eigenvalue matrix

$$\Lambda \equiv \begin{bmatrix} \lambda_1 & 0 & 0 \\ 0 & \lambda_2 & 0 \\ 0 & 0 & \lambda_3 \end{bmatrix} \qquad 1.59b$$

then

$$\Gamma^{-1/2} = U\,\Lambda^{-1/2}\,U^{+} \qquad 1.60a$$

with

$$\Lambda^{-1/2} = \begin{bmatrix} \pm\lambda_1^{-1/2} & 0 & 0 \\ 0 & \pm\,\lambda_2^{-1/2} & 0 \\ 0 & 0 & \pm\,\lambda_3^{-1/2} \end{bmatrix} \; . \qquad 1.60b$$

There is apparently an 8-fold freedom in the choice of signs in 1.60b. However, setting all nuclear displacements equal to zero, and writing

$$\underset{\sim}{r}^{\alpha} = \underset{\sim}{c}^{\alpha} \equiv \hat{\underset{\sim}{f}}{}'_j\, a_j^{\alpha} \qquad 1.61$$

for some right-handed orthonormal triple $\{\hat{\underset{\sim}{f}}{}'_i\}$, we find that, in order for the $\{\hat{\underset{\sim}{f}}_i\}$ calculated <u>via</u> 1.58 to be identical with the initial set $\{\hat{\underset{\sim}{f}}{}'_i\}$ the sign choice must be such that $\Gamma^{-1/2}$ is the 'positive-definite' square root of Γ^{-1} (recall that $\{\hat{\underset{\sim}{e}}_i\}$ is a principal-axis frame for the static model)

$$\Lambda^{-1/2} = \begin{bmatrix} +\lambda_1^{-1/2} & 0 & 0 \\ 0 & +\lambda_2^{-1/2} & 0 \\ 0 & 0 & +\lambda_3^{-1/2} \end{bmatrix} . \qquad 1.62$$

This sign convention is therefore adopted for finite nuclear displacements.

At this point, it can be seen how the possibility of a left-handed Eckart frame ((det C) = −1) arises naturally. Consider an arbitrary nuclear configuration $\{\underset{\sim}{r}^{\alpha}\}$, and calculate the determinant of the direction-cosine matrix C

$$(\det C) = (\det(\hat{\underset{\sim}{\ell}}\cdot\hat{\underset{\sim}{f}})) = \det(\hat{\underset{\sim}{\ell}}\cdot\underset{\sim}{F})\det(\Gamma^{-1/2}) = \det(\hat{\underset{\sim}{\ell}}\cdot\underset{\sim}{F})(\det \Lambda^{-1/2}) . \qquad 1.63$$

As it has already been stipulated that $(\det \Lambda^{-1/2}) > 0$, i.e., $\Lambda^{-1/2}$ is positive-definite, $(\det C) < 0$ as soon as $(\det(\hat{\underset{\sim}{\ell}}\cdot\underset{\sim}{F})) < 0$, in which case the Eckart frame is left-handed.

Hence, according to the prescription 1.58 we would expect to find det C = −1 in some region of nuclear configuration space, probably involving large-amplitude displacements from an initial configuration such as 1.61. In fact, the anticipated changeover from det C = +1 to det C = −1 as the nuclei move away from their equilibrium positions $\{\underset{\sim}{c}^{\alpha}\}$ has apparently been observed in computer calculations by Meyer and Redding [13], who refer to this phenomenon as 'Eckart frame inversion'. It demonstrates that the requirements of a right-handed Eckart frame and positive-definiteness of $\Gamma^{-1/2}$ are not compatible throughout all nuclear configuration space. This problem was not considered by Eckart, who assumed that $\{\hat{\underset{\sim}{f}}_i\}$ would be right-handed at all times, and that all nuclear displacements were in any case 'small'. The singular points at which $\det(\hat{\underset{\sim}{\ell}}\cdot\underset{\sim}{F}) = 0$ correspond to linear dependence of the Eckart vectors $\{\underset{\sim}{F}_i\}$ [39].

These results show that in general we should be prepared to consider both right- and left-handed Eckart frames together i.e., $C \in O(3)$. Such a point of view is central to the consistent treatment of the parity of molecular wavefuntions (cf. Chapter 2 and Appendix 2).

3) Following Sørensen [5], let us examine the general relation between molecule-fixed cartesian components of nuclear displacements $\{d_i^\alpha\}$ and the internal coordinates $\{q_\lambda;\ \lambda = 1,\ldots,3N-6\}$. For quasi-rigid molecules undergoing small-amplitude vibrations, we assume that an expansion about the equilibrium configuration

$$q_\lambda = \sum_\alpha \left(\frac{\partial q_\lambda}{\partial r_j^\alpha}\right)_0 d_j^\alpha + \frac{1}{2}\sum_{\alpha\alpha'} \left(\frac{\partial^2 q_\lambda}{\partial r_j^\alpha \partial r_j^{\alpha'}}\right)_0 d_j^\alpha d_{j'}^{\alpha'} + \cdots \qquad 1.64a$$

is convergent, with inverse transformation

$$d_j^\alpha = \sum_\alpha \left(\frac{\partial r_j^\alpha}{\partial q_\lambda}\right)_0 q_\lambda + \frac{1}{2}\sum_{\lambda\lambda'} \left(\frac{\partial^2 r_j^\alpha}{\partial q_\lambda \partial q_{\lambda'}}\right)_0 q_\lambda q_{\lambda'} + \cdots \qquad 1.64b$$

Comparison with the expansions (1.30b)

$$s_{\lambda,\alpha j} = \left(\frac{\partial q_\lambda}{\partial r_j^\alpha}\right)_0 + \sum_{\alpha'} \left(\frac{\partial^2 q_\lambda}{\partial r_j^\alpha \partial r_{j'}^{\alpha'}}\right)_0 d_{j'}^{\alpha'} \qquad 1.65a$$

and (1.13c)

$$t_{\alpha j,\lambda} = \left(\frac{\partial r_j^\alpha}{\partial q_\lambda}\right)_0 + \sum_{\lambda'} \left(\frac{\partial^2 r_j^\alpha}{\partial q_\lambda \partial q_{\lambda'}}\right)_0 q_{\lambda'} \qquad 1.65b$$

shows that the general coordinate transformation 1.64 can be written

$$q_\lambda = \sum_\alpha s^0_{\lambda,\alpha j} d_j^\alpha + \frac{1}{2}\sum_{\alpha\alpha'} \left(\frac{\partial s^0_{\lambda,\alpha j}}{\partial r_{j'}^{\alpha'}}\right)_0 d_j^\alpha d_{j'}^{\alpha'} + \cdots \qquad 1.66a$$

$$d_j^\alpha = \sum_\lambda t^0_{\alpha j,\lambda} q_\lambda + \frac{1}{2}\sum_{\lambda\lambda'} \left(\frac{\partial t_{\alpha j,\lambda}}{\partial q_{\lambda'}}\right)_0 q_\lambda q_{\lambda'} + \cdots\ . \qquad 1.66b$$

These general nonlinear relations correspond to the use of curvilinear coordinates q_λ. It is possible to develop the molecular Hamiltonian in terms of such coordinates, using the Eckart constraints, but the resulting theory is rather complicated [40]. It proves much simpler to define rectilinear coordinates by the truncated expansion 1.66b

$$d_j^\alpha = \sum_\lambda t^0_{\alpha j,\lambda} q_\lambda \qquad 1.67$$

so that the nuclear displacements are strictly linearly related to the values of the internal coordinates. Theories formulated in terms of rectilinear coordinates q_λ and corresponding curvilinear coordinates $\bar{q}_\lambda$ (which may have a well-defined geometrical significance) are identical for infinitesimal nuclear displacements, where a harmonic approximation to the potential energy can be strictly valid. One motivation for the introduction of curvilinear coordinates is the treatment of nuclear motions having amplitudes greater than infinitesimal, and a useful practical approach for large-amplitude motions is to embody the curvilinear coordinates directly into a semi-rigid molecular model, treating remaining degrees of freedom as rectilinear coordinates representing small-amplitude displacements from the semi-rigid model [5] (§1.4). The linear relation 1.67 simplifies both the formal development of the molecular Hamiltonian and the treatment of symmetry properties (cf. Chapter 2), and the use of 3N-6 rectilinear coordinates is appropriate for rigid molecules (note, however, that the mass-dependence of the Eckart conditions means that geometrically-defined coordinates are most suitable when comparing the properties of isotopic species [41]).

Defining the zeroth order rotational t-vector (cf. 1.13b)

$$\underset{\sim}{t}^0_{\alpha,Ri} \equiv \hat{\underset{\sim}{f}}_i \wedge \underset{\approx}{c}^{\alpha} , \qquad 1.68$$

we have the relation between $\underset{\sim}{t}^0_{\alpha,Ri}$ and the constraint vector (cf. 1.51)

$$m_\alpha \underset{\sim}{t}^0_{\alpha,Ri} = \underset{\sim}{c}_{i,\alpha} \quad . \qquad 1.69$$

It must be stressed that the use of rectilinear coordinates does <u>not</u> mean that the expansion for the $\underset{\sim}{s}_{\lambda,\alpha}$ (which appear in the kinetic energy) can be truncated in the same fashion. This is because we require that the vibrational s-vectors be subject to the generalized Malhiot-Ferigle conditions 1.28. Following Sørensen it is found that vibrational s-vectors satisfying the appropriate orthogonality relations can be formed by adding small contributions from rotational s-vectors

$$\underset{\sim}{s}_{\lambda,\alpha} = \underset{\sim}{s}^0_{\lambda,\alpha} - \sum_{\lambda'} q_{\lambda'} \zeta^{i'}_{\lambda'\lambda} \underset{\sim}{s}_{Ri',\alpha} \qquad 1.70$$

where we have defined the <u>Coriolis coupling constants</u>

$$\zeta^{i}_{\lambda'\lambda} \equiv \varepsilon_{ijk} \sum_{\alpha} t^0_{\alpha j,\lambda'} s^0_{\lambda,\alpha k} \qquad 1.71$$

for any set of rectilinear coordinates, and assume that the usual zeroth order Malhiot-Ferigle conditions are satisfied [29]

$$\sum_{\alpha} \underset{\sim}{s}^0_{\lambda,\alpha} \cdot \underset{\sim}{t}^0_{\alpha,Ri} = 0 \qquad 1.72$$

(the <u>zeroth order</u> vibrational s-vectors are the usual Wilson s-vectors [6]).

We have now found all the s-vectors appropriate for nuclear motion in a quasi-rigid molecule, and can proceed directly to the molecular Hamiltonian in the next section.

1.3. The Wilson-Howard-Watson Hamiltonian

The component blocks of the generalized G-matrix (ignoring translation terms, which we assume separated) can now be obtained:

1) μ-tensor

The pure rotational part of the kinetic energy is

$$2T_{rot} \equiv \mu_{ii'}J_iJ_{i'} \tag{1.73}$$

where $$J_i \equiv (\partial T/\partial \omega_i) \tag{1.74a}$$

are components of the total molecular angular momentum projected onto the molecule-fixed frame, and

$$[\mathbf{G}]_{rot} \equiv \mu = \boldsymbol{\eta}\, \mathbf{I}^0\, \tilde{\eta} \tag{1.74b}$$

is the inverse of the effective inertia tensor. Recalling that

$$\eta = \mathbf{A}^{-1} \;; \quad A_{ii'} = I^0_{ii'} + \varepsilon_{ijk}\varepsilon_{i'j'k} \sum m_\alpha a^\alpha_j d^\alpha_{j'} \tag{1.75a}$$

$$\equiv I^0_{ii'} + \frac{1}{2}\, a_{ii'} \tag{1.75b}$$

i.e., $$\mathbf{A} = \mathbf{I}^0 + \frac{1}{2}\,\mathbf{a} \tag{1.75c}$$

and using the matrix identity

$$\mathbf{A}^{-1} \equiv (\mathbf{I}^0)^{-1}(\mathbf{1} - \frac{1}{2}\,\mathbf{a}\mathbf{A}^{-1}) \tag{1.76}$$

we have

$$\boldsymbol{\eta} = \sum_{n=0,1\ldots}^{\infty} \boldsymbol{\mu}^0(-\frac{1}{2}\,\mathbf{a}\boldsymbol{\mu}^0)^n \tag{1.77a}$$

where the inverse of the equilibrium inertia tensor is

$$\boldsymbol{\mu}^0 \equiv (\mathbf{I}^0)^{-1} \;. \tag{1.77b}$$

Noting that $\boldsymbol{\eta}$ is a symmetric matrix, we derive the expansion of the tensor $\boldsymbol{\mu}$ about $\boldsymbol{\mu}^0$ in powers of the vibrational coordinates

$$\mu = \sum_{n=0,1\ldots}^{\infty} (n+1)\mu^0(-\frac{1}{2}a\mu^0)^{-1} \qquad 1.78$$

(cf. [26], equation 61; also [42]).

2) Coriolis Coupling

The part of the kinetic energy involving the coupling of vibrational and angular momenta is

$$T_{vib-rot} = \sum J_i G_{i\lambda} P_\lambda \qquad 1.79$$

where
$$P_\lambda \equiv (\partial T/\partial \dot{q}_\lambda) \qquad 1.80a$$

and
$$G_{i\lambda} = \eta_{ii'} \sum_\alpha \underset{\sim}{t}^0_{\alpha,Ri'} \cdot \underset{\sim}{s}_{\lambda,\alpha} \ . \qquad 1.80b$$

These G-matrix elements vanish at equilibrium ($q_\lambda = 0$)

$$G^0_{i\lambda} = \eta_{ii'} \sum \underset{\sim}{t}^0_{\alpha,Ri'} \cdot \underset{\sim}{s}^0_{\lambda,\alpha} = 0 \qquad 1.81$$

owing to the zeroth-order Malhiot-Ferigle (Eckart) conditions, so that the coupling terms are in fact second order in the vibrational coordinates (q_λ, p_λ) [43]. Hence, the Eckart conditions ensure that the Coriolis coupling vanishes to first order, as stated above.

Introducing vibrational s-vectors appropriate for rectilinear coordinates, we have

$$G_{i\lambda} = -\sum \mu_{ij}\zeta^j_{\lambda'\lambda} q_{\lambda'} \qquad 1.82$$

so that defining the 'vibrational angular momentum'

$$\pi_i \equiv \sum \zeta^i_{\lambda'\lambda}\, q_{\lambda'} p_\lambda \qquad 1.83$$

the coupling term becomes

$$T_{vib-rot} = -\mu_{ii'} J_i \pi_{i'} \ . \qquad 1.84$$

As emphasized by Watson [26], the components $\{\pi_i\}$ of the vibrational angular momentum do not in general obey the usual commutation (Poisson bracket) relations.

3) The Vibrational G-matrix

Vibrational G-matrix elements are

$$G_{\lambda\lambda'} = \sum m_\alpha^{-1} \underset{\sim}{s}_{\lambda,\alpha} \cdot \underset{\sim}{s}_{\lambda',\alpha} \qquad 1.85$$

and will in general depend upon the internal coordinates q_λ. In the special case of rectilinear coordinates, we have

$$\underset{\sim}{s}_{\lambda,\alpha} = \underset{\sim}{s}^0_{\lambda,\alpha} - \sum_\alpha q_{\lambda'} \zeta^i_{\lambda'\lambda} \underset{\sim}{s}_{Ri,\alpha} \qquad 1.70$$

In the usual theory of molecular vibrations, only the constant part G^0 of the vibrational G-matrix is considered explicitly,

$$G^0_{\lambda\lambda'} \equiv \sum m_\alpha^{-1} \underset{\sim}{s}^0_{\lambda,\alpha} \cdot \underset{\sim}{s}^0_{\lambda',\alpha} \qquad 1.86$$

the remainder being treated later, if necessary, as a perturbation. It is also assumed that the expansion of the molecular potential energy in internal coordinates

$$V(q_\lambda) = V_0 + \sum \left(\frac{\partial V}{\partial q_\lambda}\right)_0 q_\lambda + \frac{1}{2} \sum \left(\frac{\partial^2 V}{\partial q_\lambda \partial q_{\lambda'}}\right)_0 q_\lambda q_{\lambda'} \qquad 1.87$$

can initially be truncated after the second term, and the term linear in q ignored (stable reference structure). We then seek a linear transformation with constant coefficients L to a set of <u>normal coordinates</u> $\{Q_\lambda\}$

$$q_\lambda = \sum L_{\lambda\lambda'} Q_{\lambda'} \qquad 1.88$$

such that the kinetic energy matrix $\mathbf{G}^0$ and the quadratic force constant matrix $\mathbf{F}$ $\left(F_{\lambda\lambda'} \equiv (\partial^2 V/\partial q_\lambda \partial q_{\lambda'})_0\right)$ are simultaneously diagonalized

$$\tilde{\mathbf{L}}\ \mathbf{F}\ \mathbf{L} = \Lambda \qquad 1.89a$$

$$\mathbf{L}^{-1}\ \mathbf{G}^0\ \tilde{\mathbf{L}}^{-1} = \mathbf{1} \qquad 1.89b$$

where Λ is the matrix of vibrational frequencies. This leads to the familiar equations of the Wilson FG-method

$$\mathbf{L}^{-1}(\mathbf{G}^0\mathbf{F})\mathbf{L} = \Lambda \qquad 1.90a$$

$$\tilde{\mathbf{L}}(\mathbf{F}\mathbf{G}^0)\tilde{\mathbf{L}}^{-1} = \Lambda \qquad 1.90b$$

Denoting the s- and t-vectors appropriate for normal coordinates by $\underset{\sim}{\sigma}$, $\underset{\sim}{\tau}$ respectively, we introduce ℓ-vectors <u>via</u>

$$\underset{\sim}{\sigma}^{0}_{\lambda,\alpha} \equiv m_{\alpha}^{1/2} \underset{\sim}{\ell}_{\lambda,\alpha} \tag{1.91a}$$

$$\underset{\sim}{\tau}^{0}_{\alpha,\lambda} \equiv m_{\alpha}^{-1/2} \underset{\sim}{\ell}_{\lambda,\alpha} \tag{1.91b}$$

with

$$\sum_{\alpha} \underset{\sim}{\ell}_{\lambda,\alpha} \cdot \underset{\sim}{\ell}_{\lambda',\alpha} = \delta_{\lambda\lambda'} \quad . \tag{1.92}$$

Use of the ℓ-vectors corresponds to an orthogonal transformation to molecular coordinates from mass-weighted cartesian coordinates. We therefore obtain the usual definition of Coriolis coupling coefficents in terms of rectilinear normal coordinates

$$\zeta^{i}_{\lambda\lambda'} = \epsilon_{ijk} \sum_{\alpha} \ell_{\lambda,\alpha j} \ell_{\lambda',\alpha k} \quad . \tag{1.93}$$

Thus, assuming rectilinear normal coordinates, and drawing upon all the above results, the classical Hamiltonian form of the molecular energy excluding translation is

$$H = \frac{1}{2} \mu_{ii'} (J_i - \pi_i)(J_{i'} - \pi_{i'}) + \frac{1}{2} \sum P^2_{\lambda} + V(Q_{\lambda}) \tag{1.94}$$

where

$$P_{\lambda} \equiv (\partial T/\partial \dot{Q}_{\lambda}) \tag{1.95}$$

is the momentum conjugate to the normal coordinate Q_{λ}, and all other quantities appearing in 1.94 have been dealt with above.

Defining the vector $\underset{\sim}{\mathbf{K}}$ as follows (cf. Appendix 2; also [7])

$$\mathbf{K}_i \equiv - J_i \tag{1.96}$$

($\mathbf{K}_i$ = -(det C)J_i for C ϵ O(3)), we obtain the desired classical result

$$H = \frac{1}{2} \mu_{ii'} (\mathbf{K}_i + \pi_i)(\mathbf{K}_{i'} + \pi_{i'}) + \frac{1}{2} \sum_{\lambda} P^2_{\lambda} + V(Q_{\lambda}) \quad . \tag{1.97}$$

Finally, we must consider the problem of quantizing the classical Hamiltonian 1.97; that is, we must find the corresponding operator $\hat{H}$ acting upon the space of square-integrable functions of the vibrational and rotational coordinates. Here we shall only state the end result, which is of remarkable simplicity: the quantum molecular Hamiltonian $\hat{H}$ is identical in form with the classical expression 1.97, except for the appearance of an additional mass-dependent 'quantum potential' U discovered by Watson [26]

$$\hat{H} = \frac{1}{2} \mu_{ii'}(\hat{K}_i + \hat{\pi}_i)(\hat{K}_{i'} + \hat{\pi}_{i'}) + \frac{1}{2} \sum \hat{P}_\lambda^2 + V(Q_\lambda) + U \qquad 1.98$$

where the extra term U is

$$U = -(\hbar^2/8)\mu_{ii} \ . \qquad 1.99$$

The ordering of operators in the first term of $\hat{H}$ is unimportant, since [26]

$$(\hat{K}_i + \hat{\pi}_i)\mu_{ii'} = \mu_{ii'}(\hat{K}_i + \hat{\pi}_i) \qquad 1.100$$

(recall there is an implicit summation over i).

It is intriguing to reflect that there apparently exists no simple route to this exceptionally compact result. Watson's original derivation [26] made extensive use of some rather complicated sum rules, whilst Louck's recent rederivation [7] involving a direct transformation of the nuclear kinetic energy operator to molecular coordinates is no less lengthy or difficult. Essen [44] has discussed the significance of this work from a general viewpoint (see also [45,46]).

The molecular Hamiltonian 1.98 is a very complicated object, and to make any progress we must introduce expansions for both the μ-tensor and the potential energy. Thus, retaining only the first term in 1.78, expanding the potential energy up to quadratic terms, and ignoring both U and the $\hat{\pi}_i$, the Hamiltonian becomes

$$\hat{H}^0 - V_0 = \mu_{ii'}^0 \hat{K}_i \hat{K}_{i'} + \frac{1}{2} \sum P_\lambda^2 + \frac{1}{2} \sum \omega_\lambda^2 Q_\lambda^2 \qquad 1.101a$$

$$= \sum_{i=xyz} (\hat{K}_i^2/I_i^0) + \frac{1}{2} \sum_\lambda (P_\lambda^2 + \omega_\lambda^2 Q_\lambda^2) \ . \qquad 1.101b$$

This result corresponds to our intuitive zeroth-order rigid-rotor/harmonic oscillator picture of rigid molecules [47], and is the starting point for a perturbation theory treatment of vibration-rotation energy levels [18,43].

As shown by Louck [7], the angular momentum operators $\hat{K}_i$ can be expressed directly in terms of the matrix elements C_{ij} ($\hbar = 1$)

$$\hat{K}_i = - i\varepsilon_{ijk}\tilde{C}_{js}\ \partial/\partial\tilde{C}_{ks} \qquad 1.102$$

where the components of $\hat{\underset{\sim}{K}}$ obey normal angular momentum commutation relations

$$[\hat{K}_i, \hat{K}_j] = i\varepsilon_{ijk}\hat{K}_k \qquad 1.103$$

The significance of these results is discussed in Appendix 2.

1.4 Nonrigid Molecules

In this section, it is shown that a Hamiltonian appropriate for certain types of nonrigid molecule (NRM) can be derived using a straightforward generalization of the procedure just described for rigid molecules. The key point here is the introduction of the semi-rigid molecular model; as we discuss in Chapter 3, this concept is also of central importance for the treatment of the symmetry properties of NRMs.

Let us then consider the semi-rigid molecular model (SRMM) [5,8,48,49]. The utility of this concept derives from the fact that, at least for certain classes of NRMs, the nuclear dynamics are best described in terms of the intuitive picture introduced by Sayvetz [10], in which the nuclei execute rapid, small-amplitude (vibrational) motion about an 'equilibrium' configuration that is itself performing some sort of slow, large-amplitude (internal or contortional [97]) motion, as well as undergoing overall rotation. The internal motion embodied in the specification of the SRMM may involve some form of large-amplitude bending [50], internal rotation [51-53], inversion at (for example) a nitrogen atom [54], or a more complex type of motion such as pseudorotation [55] (cf. Chapters 3 and 4). As will be seen, the SRMM underlies the description of vibrational motions in NRMs in precisely the same way that the static molecular model does in quasi-rigid molecules. In the case of NRMs, the reference structure can be embedded into an arbitrary spatial configuration of nuclei using the Eckart-Sayvetz conditions [10], which are a natural generalization of the Eckart conditions to nonrigid systems.

Formally, the SRMM is a set of triples (cf. 1.44) [48,49]

$$\mathcal{A} \equiv \{(\underset{\sim}{a}^{\alpha}(\gamma), m_{\alpha}, z_{\alpha}); \quad \alpha = 1,\ldots,N\} \qquad 1.104$$

defining N vectors $\underset{\sim}{a}^{\alpha}(\gamma)$ as functions of a set of parameters collectively denoted γ, where each $\underset{\sim}{a}^{\alpha}(\gamma)$ is associated with a nucleus of mass m_{α} and charge z_{α}. γ is a representative point in the parameter domain Γ, and we shall suppose that there are T ($\lesssim$ 3N-6) independent parameters, so that γ stands for the vector $(\gamma_1\cdots\gamma_t\cdots)$. Here, each γ_t is a coordinate, such as a torsional angle or angle of pucker, describing a particular large-amplitude motion, and each has a finite range defined in a suitable fashion [48]. It is clear that the γ_t are curvilinear coordinates in the sense of §1.3, and indeed the essence of the method considered is that the T curvilinear coordinates are treated on the same footing as the large-amplitude translation and rotation coordinates (thereby necessitating T extra constraints on the nuclear displacements), leaving 3N-6-T rectilinear coordinates to describe small-amplitude vibrations.

The 3N components of the SRMM vectors

$$a^{\alpha}(\gamma)_i \equiv \hat{\underset{\sim}{e}}_i \cdot \underset{\sim}{a}^{\alpha}(\gamma) \qquad i = x,y,z \qquad 1.105$$

are defined by introducing a coordinate frame $\{\hat{\underset{\sim}{e}}_i\}$ into the SRMM, where $\{\hat{\underset{\sim}{e}}_i\}$ may be a principal-axis frame for a particular value of γ. It is convenient to require that the centre of mass condition

$$\sum m_{\alpha} a^{\alpha}(\gamma)_i = 0 \qquad \text{for all } \gamma; \quad i = x,y,z \qquad 1.106$$

be satisfied.

The actual definition of the $a^{\alpha}(\gamma)_i$ as functions of the parameters γ requires careful consideration when we come to the problem of formulating a multiplication rule in the symmetry group of the molecular model (Chapter 3; [49,56]). For the moment, we mention three possibilities [5,50]:

a) Principal-axis method: this requires that the coordinate frame $\{\hat{\underset{\sim}{e}}_i\}$ be a principal-axis frame for all values taken by the parameters

$$\sum_{\alpha} m_{\alpha} a^{\alpha}(\gamma)_i a^{\alpha}(\gamma)_{i'} = 0 \qquad \text{for all } \gamma;\ i \neq i' . \qquad 1.107a$$

This choice is convenient for systems with highly symmetric rotors, and obviously simplifies the zeroth-order treatment of rotational motion.

b) Internal-axis method: this requires that an equation of the form [50]

$$\varepsilon_{ijk} \sum_{\alpha} m_\alpha a^\alpha(\gamma)_j \left(\frac{\partial a^\alpha(\gamma)_k}{\partial \gamma_t}\right) = 0 \qquad \text{for all } \gamma \qquad 1.107b$$

be satisfied, which eliminates coupling between internal and rotational motions in zeroth order. There are, however, some difficulties associated with the general use of this method. First, it is not clear that equation 1.107b is in general integrable. Although it might be satisfied locally, globally it may be possible to induce a finite rotation of the molecular model by a finite variation of the parameters, thereby introducing a path-dependence or multi-valuedness into our description. Second, as pointed out by Sørensen [5], although the condition 1.107b removes rotation/internal-motion coupling to zeroth order, any <u>effective</u> rotation/internal-motion Hamiltonian obtained by projecting out excited vibrational states [22] will in general contain such coupling terms anyway, arising from second order contributions, so that the original simplification is lost. Third, as noted by Gilles and Philippot [56] (see also [23,49]), the condition 1.107b can introduce unnecessary complications in the description of the symmetry properties of even quite simple NRMs (cf. Chapter 3).

c) Geometric method: here, we adopt an analytic or geometric definition of the $a^\alpha(\gamma)_i$ as functions of γ, regardless of dynamical constraints such as a) or b). This method is most useful when treating the symmetry properties of arbitrary NRMs.

For molecules with nonrigidity described by the Sayvetz picture, the coordinate transformation 1.43 generalizes to

$$R_i^\alpha - R_i = C_{ij}(a^\alpha(\gamma)_j + d_j^\alpha(q_\lambda)) \qquad 1.108a$$

i.e.,

$$\underset{\sim}{r}^\alpha = \hat{\underset{\sim}{f}}_i(a^\alpha(\gamma)_j + d_j^\alpha(q_\lambda)) \qquad 1.108b$$

where the reference positions of the nuclei now depend upon the dynamical variables γ (corresponding to the SRMM parameters), and the nuclear displacements $\{d_j^\alpha\}$ depend upon 3N-6-T rectilinear coordinates $\{q_\lambda;\lambda = 1,\ldots,3N\text{-}6\text{-}T\}$. In what follows, our equations refer to SRMMs having a single large-amplitude coordinate only. Formally, there is no problem in extending the treatment to molecules with any number of curvilinear coordinates [8]; in practice, however, the presence of more than one parameter in the SRMM can pose quite a difficult dynamical problem (e.g. [57]).

In our treatment of the classical Hamiltonian for NRMs, we proceed exactly as for rigid molecules. Thus, introducing t-vectors, the velocity of nucleus α is

$$d/dt_{lab}\, \underset{\sim}{r}^{\alpha} = \underset{\sim}{\omega}\wedge\underset{\sim}{r}^{\alpha} + \underset{\sim}{t}_{\alpha,\gamma}\,\dot{\gamma} + \sum_{\lambda} \underset{\sim}{t}_{\alpha,\lambda}\dot{q}_{\lambda} \qquad 1.109$$

where
$$\underset{\sim}{t}_{\alpha,\gamma} \equiv (\partial \underset{\sim}{r}^{\alpha}/\partial\gamma) \qquad 1.110a$$

and
$$\underset{\sim}{t}_{\alpha,\lambda} \equiv (\partial \underset{\sim}{r}^{\alpha}/\partial q_{\lambda}) \qquad \lambda = 1,\ldots,3N-6\ . \qquad 1.110b$$

Before deriving s-vectors for NRMs, it is necessary to consider the extra constraint associated with the internal parameter γ. Following our previous discussion of the rotational Eckart constraints we introduce the constraint vector (cf. 1.69)

$$\underset{\sim}{c}_{\gamma,\alpha} \equiv m_{\alpha}\underset{\sim}{t}^{0}_{\alpha,\gamma} = m_{\alpha}(\partial \underset{\sim}{c}^{\alpha}/\partial\gamma) \qquad 1.111$$

and have (cf. 1.36)

$$\sum_{\alpha} \underset{\sim}{c}_{\gamma,\alpha} \cdot \underset{\sim}{t}^{0}_{\alpha,\lambda} = 0 \qquad \text{for all } \lambda \qquad 1.112$$

Since we assume the use of rectilinear coordinates q_{λ}, equation 1.112 can be integrated to give

$$\sum_{\alpha} \underset{\sim}{c}_{\gamma,\alpha} \cdot \underset{\sim}{d}^{\alpha} = 0 \qquad 1.113a$$

i.e.,
$$\sum_{\alpha} m_{\alpha}\, (\partial a^{\alpha}(\gamma)_j/\partial\gamma) d^{\alpha}_j = 0 \qquad 1.113b$$

which is the Eckart-Sayvetz condition [10]. The constraint 1.113 uncouples the internal large-amplitude motion from nuclear vibrations to first order in the nuclear displacements, and is therefore analogous to the rotational Eckart conditions (1.49). We shall refer to the molecule-fixed coordinate frame determined by the Eckart and Eckart-Sayvetz conditions as the Eckart-Sayvetz frame.

It is appropriate to mention here the least-squares characterization of the Eckart-Sayvetz frame obtained by Natanson and Adamov, and more recently by Jørgensen [39]. Thus, consider a particular nuclear configuration $\{\underset{\sim}{r}^{\alpha}\}$

$$\underset{\sim}{r}^{\alpha} = \bar{\underset{\sim}{c}}^{\alpha}(\bar{\gamma}) + \bar{\underset{\sim}{d}}^{\alpha} \qquad 1.114a$$

where

$$\bar{\underset{\sim}{c}}^{\alpha} \equiv \hat{\underset{\sim}{f}}{}'_j \, a^{\alpha}(\gamma)_j \qquad 1.114b$$

for some $\{\hat{\underset{\sim}{f}}{}'_i\}$ and $\bar{\gamma}$, and all the nuclear displacements $\{\bar{\underset{\sim}{d}}^{\alpha}\}$ are in some sense 'small'. Then, forming the quantity

$$\mathcal{E} \equiv \sum m_{\alpha} \, \bar{\underset{\sim}{d}}^{\alpha} \cdot \bar{\underset{\sim}{d}}^{\alpha} \qquad 1.115$$

it is simple to prove that varying both the orientation of the coordinate frame $\{\hat{\underset{\sim}{f}}{}'\}$ and the parameter $\bar{\gamma}$ until $\mathcal{E}$ is a minimum uniquely defines the Eckart-Sayvetz frame, together with corresponding value of γ.

The result is important for two reasons. First, for NRMs the least-squares criterion based on 1.115 avoids explicit consideration of the Sayvetz constraint 1.113, which would involve differentiation with respect to γ. Second, and more important for our purposes, we note that the quantity $\mathcal{E}$ is <u>invariant</u> under the transformation

$$d_i^{\alpha} \rightarrow \sum \; S_{\alpha\beta} R_{ij} d_j^{\beta} \qquad 1.116$$

where R is an arbitrary rotation matrix (Appendix 1), and S is an N by N permutation matrix that interchanges identical nuclei. The implications of this invariance for the theory of molecular symmetry are discussed in Chapters 2 and 3.

Returning to the NRM Hamiltonian, we introduce an index p taking values x,y,z,γ, and define the 4 by 4 matrix **A**

$$A_{pp'} \equiv \sum_{\alpha} m_{\alpha} \underset{\sim}{t}{}^{0}_{\alpha,p} \cdot \underset{\sim}{t}{}^{0}_{\alpha,p'} \qquad 1.117$$

which is a generalization of the 3 by 3 **A** matrix for rigid molecules (1.39). With (cf. 1.39)

$$\eta \equiv \mathbf{A}^{-1} \qquad 1.118$$

the rotational/internal-motion s-vectors are [5] (cf. 1.38)

$$\underset{\sim}{s}_{p,\alpha} = \eta_{pp'} m_\alpha \underset{\sim}{t}_{\alpha,p'} \qquad p = x,y,z,\gamma \tag{1.119}$$

Introducing a γ-dependent transformation to normal coordinates

$$d_j^\alpha = \sum_\lambda m_\alpha^{-1/2}\, \ell(\gamma)_{\lambda,\alpha j} Q_\lambda \quad , \tag{1.120a}$$

where
$$\sum_\alpha \ell(\gamma)_{\lambda,\alpha j}\ell(\gamma)_{\lambda',\alpha j} = \delta_{\lambda\lambda'} \quad \text{for all } \gamma \tag{1.120b}$$

and the Coriolis coupling coefficients $\zeta^p_{\lambda\lambda'}$ are

$$\zeta(\gamma)^i_{\lambda\lambda'} \equiv \varepsilon_{ijk} \sum m_\alpha \ell(\gamma)_{\lambda,\alpha j}\ell(\gamma)_{\lambda',\alpha k} \tag{1.121a}$$

$$\zeta(\gamma)^\gamma_{\lambda\lambda'} \equiv \sum_\alpha (\partial\ell(\gamma)_{\lambda,\alpha j}/\partial\gamma)\ell(\gamma)_{\lambda',\alpha j} = -\zeta(\gamma)^\gamma_{\lambda\lambda'} \, , \tag{1.121b}$$

vibrational s-vectors are

$$\underset{\sim}{s}_{\lambda,\alpha} = \underset{\sim}{s}^0_{\lambda,\alpha} - \sum_{\lambda'} Q_{\lambda'}\zeta^i_{\lambda'\lambda}\underset{\sim}{s}_{Ri,\alpha} - \sum_{\lambda'} Q_{\lambda'}\zeta^\gamma_{\lambda'\lambda}\, \underset{\sim}{s}_{\gamma,\alpha} \tag{1.122}$$

and satisfy the required orthogonality relation (cf. 1.26)

$$\sum_\alpha \underset{\sim}{s}_{\lambda,\alpha} \cdot \underset{\sim}{t}_{\alpha,p} = 0 \quad \text{for all } \lambda,\ p \ . \tag{1.123}$$

Finally, defining the quantities

(cf. 1.83)
$$\pi_p \equiv \sum \zeta^p_{\lambda\lambda'} Q_\lambda P_{\lambda'} \tag{1.124a}$$

$$P_\lambda = (\partial T/\partial \dot{Q}_\lambda) \tag{1.124b}$$

(cf. 1.74)
$$\mu \equiv \eta\, I\, \tilde{\eta} \tag{1.124c}$$

(cf. 1.52)
$$I^0_{\lambda\lambda'} \equiv \sum m_\alpha^{-1}\, \underset{\sim}{c}_{p,\alpha} \cdot \underset{\sim}{c}_{p',\alpha} \tag{1.124d}$$

(cf. 1.96)
$$\mathbf{K}_i \equiv -J_i = -\ \partial T/\partial\omega_i \tag{1.124e}$$

$$\mathbf{K}_\gamma \equiv -P_\gamma \equiv -(\partial T/\partial\dot{\gamma}) \tag{1.124f}$$

the classical Hamiltonian for an NRM with one large-amplitude curvilinear coordinate takes the form

$$H = \frac{1}{2}\, \mu_{pp'}(\mathbf{K}_p + \pi_p)(\mathbf{K}_{p'} + \pi_{p'}) + \frac{1}{2} \sum_{\lambda=1}^{3N-7} P_\lambda^2 + V(Q_\lambda;\gamma) \tag{1.125}$$

which is completely analogous to the rigid molecule Hamiltonian 1.97. We note that the 4-dimensional tensor I^0 generalizes the equilibrium inertia tensor of the static molecular model (1.52) to NRMs. However, I^0 is in general γ-dependent. Also, factorization of the μ-tensor occurs quite generally for NRMs (1.124c) [5]. This result had previously been derived for the particular case of quasi-linear molecules [59].

Quantization of the classical NRM Hamiltonian 1.125 does not lead to such a compact expression as that obtained for rigid molecules (1.98), since the μ-tensor is in general γ-dependent. Some simplification is possible, however, in particular cases [54].

Expanding the potential energy $V(q_\lambda;\gamma)$ in internal coordinates q_λ, we have

$$V(q_\lambda;\gamma) = V_0(\gamma) + \sum \left(\frac{\partial V}{\partial q_\lambda}\right)_0 q_\lambda + \frac{1}{2} \sum \left(\frac{\partial^2 V}{\partial q_\lambda \partial q_{\lambda'}}\right)_0 q_\lambda q_{\lambda'} + \cdots \qquad 1.126$$

where the zeroth order quantity $V_0(\gamma)$ defines a potential characterizing the large-amplitude motion e.g. a barrier hindering internal rotation. The linear term in the expansion of V does not necessarily vanish: this is because an SRMM in which bond lengths etc. are held constant does not in general define an equilibrium structure for all values of the parameter γ. However, it is in principle possible to define the SRMM such that the linear term in 1.126 vanishes for all γ, and this procedure is known as <u>structure relaxation</u> [60]. All coefficients in the expansion of the potential energy are in general γ-dependent, so that setting up the transformation to normal coordinates and the vibrational kinetic energy in internal coordinates leads to a <u>parameter dependent</u> FG-formalism, which is a generalization of the usual theory for rigid molecules [52, 183].

At this point, we have completed an outline of the orthodox approach to the dynamics of NRMs. Before going on to a study of molecular symmetry properties in Chapters 2 and 3, we pause in the next section to examine some of the more radical approaches that have been applied, or have yet to be applied, to systems whose nonrigidity is so extreme that that the conventional picture based on the SRMM is inapplicable.

1.5 Other Approaches to NRM Dynamics

Let us now consider some systems for which the notion of a semi-rigid molecular model, used previously as a basis for describing NRM dynamics, is inapplicable [61].

Van der Waals clusters: for example, a weakly-bound cluster of several inert gas atoms, such as might be formed in a supersonic nozzle beam. Apart from the intrinsic interest of the dynamics of such novel few-body systems, a knowledge of the properties of small clusters is essential for a microscopic theory of homogeneous nucleation [62]. There, we seek to determine the size of the 'critical' nucleus for condensation from the vapour, and must be able to evaluate partition functions for clusters of increasing size; this in turn requires a knowledge of the pattern of energy levels of such systems, and it is clear that the usual rigid rotor/harmonic oscillator model is inadequate here. In fact, there seems to be no definition of a semi-rigid molecular model appropriate for these clusters, since it is likely that all identical particles exchange positions on the time-scale of interest. Some other approach to the dynamics is therefore required.

Clusters of metal atoms: for example, Li_3. Several calculations [63,64] support the view that the ground electronic states of alkali metal clusters such as Li_3 or Na_3 have extremely flat, multi-minima surfaces, with the additional complication of strong vibronic (dynamic Jahn-Teller) coupling induced by a conical intersection at the equilateral configuration [64a]. Thus, even the lowest vibrational states behave as very floppy molecules, with no straightforward separation of large-amplitude internal motions and small-amplitude vibrations.

PF_5: the nonrigidity of the PF_5 molecule is easily visualized in terms of the Berry pseudorotation process [65], which is supported by available experimental evidence [66]. Difficulties here are associated with the introduction of the SRMM needed to describe a pseudorotating reference structure. It appears that there are 'global' problems of the type mentioned in §1.3, in that coupling of internal motion and overall rotation leads to an infinite-valued description of a given nuclear configuration ([67]; see also [56]). This difficulty does not arise for the pseudorotation of XPY_4-type molecules with a heavy pivotal ligand X [55], where the SRMM depends upon a single parameter.

We shall now briefly describe some approaches that have been proposed in response to these difficult problems.

Perturbative/Q-group method [34,68-70]

In this approach, a detailed description of NRM dynamics in terms of the continuous variation of some reference structure with a set of parameters is avoided by considering a finite set of reference structures, where each has a localized rigid molecule rovibronic wavefunction associated with it. The molecular Hamiltonian is then diagonalized in the rigid molecule basis, so that the initially degenerate set of rovibronic energies corresponding to the localized structure is split by quantum-mechanical tunneling. However, no matrix elements are actually evaluated explicitly, since that is, at present, a prohibitively difficult task [34]; rather, a parametrized splitting of rigid molecule levels is obtained, where the number of independent parameters is determined group-theoretically. Thus, the splitting pattern reflects the assumed structure of the Q-group, which is a supergroup of the equilibrium configuration point group generated by certain direct transformations between reference structures [68]. This method has been applied in detail to the PF_5 problem [69], and splitting patterns and line intensities associated with all possible modes of internal rearrangement have been obtained [70]. We note also the application to internal motions in XeF_6 [71].

Use of rotational constraints other than the Eckart Condition

As we have seen, use of the Eckart or Eckart-Sayvetz constraints to define the molecule-fixed axes is intimately associated with the introduction of an equilibrium structure $\{a_i^\alpha\}$. It is possible to define a molecule-fixed frame in a way that does not depend on the idea of a reference configuration [72], and which is therefore not implicitly restricted to the domain of small-amplitude motion. We consider one such possibility, namely use of the principal-axis frame [2].

Here, the molecule-fixed components $\{r_j^\alpha\}$ are required to satisfy the relation

$$Q_{jj'} = \sum m_\alpha r_j^\alpha r_{j'}^\alpha = 0 \qquad j \neq j' \qquad 1.127$$

at all times, so that $\{\hat{\underset{\sim}{f}}_i\}$ diagonalizes the mass quadrupole tensor **Q**. A calculation of the molecular Hamiltonian can then be carried through as before, starting from the coordinate transformation 1.2 and making appropriate use of the constraint 1.127 [8]. When this is done, the following interesting conclusions emerge: a) Use of the principal-axis frame does not minimize Coriolis interactions [5]. In other words, coupling of internal motions with overall rotation is in no sense small; indeed, we would not expect this to be the case. However, this has the consequence that, although the molecule-fixed frame is always well-defined (except for singular regions associated with linear or symmetric top configurations), there is no useful zeroth order vibration/rotation separation, as with the Eckart constraints. b) Intriguingly, use of the principal-axis constraints leads to the appearance of the so-called 'fluid' moments of inertia in the rotational part of the Hamiltonian [2]. In molecular physics, this is known as the Eckart paradox [3,45,73,74]. As pointed out by various authors, it is possible to effect a renormalization of the moments of inertia so as to yield 'rigid body' values [2,74].

The terms 'fluid' and 'rigid body' moments of inertia arise in the context of the theory of rotational bands in nuclear spectra [75], and the use of a principal-axis frame is standard in the general theory of collective motion in nuclei [2]. We conclude this chapter by mentioning two approaches that are in fact motivated by a 'nuclear' picture of weakly-bound clusters.

The Nuclear Picture

Consider the following well-defined few-body problem: given a set of N particles (identical for the sake of argument) interacting _via_ known (pairwise, central) potentials. What are the energy levels of the system? Stated thus, we might hope to be able to gain qualitative or even quantitative insight into the cluster problem by invoking techniques that have proved useful in the corresponding few nucleon problem [76,77].

a) Hyperspherical Harmonics

In this method, the N-body kinetic energy (minus translation) is expressed in terms of a single hyperradius (dimension length) and a generalized angular momentum operator in 3N-4 angular variables, which can be chosen in various ways [78]. The eigenfunctions of the angular portion of the kinetic energy are the hyperspherical harmonics: these are functions of the 3N-4 angles, and generalize the usual spherical harmonics (which are actually appropriate for the N = 2 body problem). The eigenfunctions of the complete Hamiltonian can be expanded in terms of the hyperspherical harmonics, where the hyperradius-dependent coefficients are determined by an infinite set of coupled differential equations (generalized partial wave expansion). Introduction of a truncated basis yields an approximate energy spectrum. In nuclear problems, it has been found that the partial wave expansion may converge quite rapidly [76].

The method has been applied in variational calculations of the ground state energies of $(He)_3$ clusters [79], and extensions to other systems are readily envisaged (for example, we note Wallace's recent treatment of large-amplitude motions in H_2O [80]).

It should particularly be recognized that it is not necessary to introduce a body-fixed frame in this method, although the theory can be developed for a principal-axis frame [2], which corresponds to a particular choice of angular variables.

b) The Oscillator Model [77]

In this approach, a limiting model of the N-body system as a cohesive cluster of particles interacting via pairwise oscillator potentials is adopted (compare the 'free particle' picture in the hyperspherical harmonic theory). Introducing standard Jacobi coordinates, the idealized Hamiltonian separates into that for 3N-3 uncoupled harmonic oscillators, each having the same frequency. As with the hyperspherical harmonics, this method provides a relatively easily handled many-particle basis set for calculations involving realistic interparticle potentials [77]. Moreover, it is again not necessary to introduce a rotating frame.

In the first application of this model to molecular problems, Berry et al. have considered the correlation between states in the harmonic oscillator limit and rigid molecule (or conventional NRM) states, in the hope of obtaining a qualitative picture of the energy level patterns of highly nonrigid systems [61,81,82].

Thus, the oscillator states are classified according to the group chain

$$U(3N - 3) \supset U(3) \times U(N - 1) \supset O^{\ell}(3) \times S_N \qquad 1.128$$

where U(3N-3) is the unitary symmetry group of the 3N-3 dimensional harmonic oscillator, U(3) is the subgroup of U(3N-3) pertaining to the 3 spatial dimensions, U(N-1) is the subgroup of U(3N-3) pertaining to the N-1 internal Jacobi vectors, $O^{\ell}(3)$ is the external rotation-inversion group, a subgroup of U(3) (cf. Appendices 1 and 2), and S_N is the permutation group on N identical particles, a subgroup of U(N-1).

As discussed in detail in Chapter 2, the rigid molecule states are classified in the group

$$O^{\ell}(3) \times G^{\pi} \qquad 1.129$$

where G^{π} is a subgroup of S_N isomorphic (for nonplanar molecules) with some point group, the symmetry group of the chosen rigid limit. Then, inducing [83, 118,125] the rigid molecule symmetry labels Γ from G^{π} to S_N

$$\Gamma(O^{\ell}(3) \times G^{\pi}) \uparrow O^{\ell}(3) \times S_N \qquad 1.130$$

and invoking the noncrossing rule gives an unambiguous correlation of states between the rigid and nonrigid limits.

The energy level spectrum of a real system is presumed to lie somewhere along the correlation diagram between the two extremes.

Chapter 2. Symmetry Properties of Rigid Molecules

In this chapter we turn to the problem of the symmetry properties of quasirigid molecules. Apart from being a question of fundamental importance in molecular physics, this subject is of some topical interest, as the availability of tunable infrared lasers has made it possible to obtain the vibration-rotation spectra of highly symmetric molecules such as CH_4, SF_6, and CF_4 at very high resolution ($\sim 10^{-3}$ cm^{-1}) and a prerequisite for interpretation of the interesting features revealed by such measurements [83] is a clear and detailed understanding of the assignment of symmetry labels to molecular states and tensor operators.

We begin by reviewing some of the classic works in this field. The subject really begins with Wigner [84], who initiated the application of group theory to the study of molecular vibrations [6, 85]. Wigner formulated vibrational symmetry operations as permutation-rotations (perrotations [56]) of nuclear displacement vectors, and showed that these operations form a group isomorphic with the covering group of the assumed nuclear equilibrium configuration. It was thereby possible to determine the number and symmetry types of the normal modes for a given molecule, and to predict which vibrations are active in the infrared, Raman etc. It was later emphasized by Eckart [9] that Wigner's treatment of vibrational symmetry implicitly assumes a particular choice of molecule-fixed axes, and that the definition of the Eckart frame corresponds to this choice (this remark has recently been taken up by Louck and Galbraith [17]: see below).

The study of the symmetry problem was continued by (amongst others) Wilson, who gave extensive tables of the numbers and degeneracies of vibrations in polyatomics of the type $AB_nB'_{n'}$ ... [86]. Other fundamental contributions were the calculation of the nuclear spin statistical weights of molecular rovibronic states [87], and the application of group theory to determine qualitative splitting patterns of vibration-rotation levels and associated selection rules [88]. It was quite clearly recognized at this stage that, when determining the effect of a point symmetry operation on the rovibronic product function, we are actually considering the induced action of a particular permutation of identical nuclei, where the complete molecular Hamiltonian is invariant under all such permutations (§2.1). However, Wilson only classified molecular states with respect to a group of nuclear permutations isomorphic with the rotational subgroup of the point group, i.e., "those permutations of identical particles which correspond to rotations of the molecule" [88], rather than the point group itself. The reason for this is that it is not immediately obvious how to define the action of improper rotations of axes on rotational wavefunctions; indeed, several approaches have been suggested to date. Wilson did not give a solution of this problem, and was therefore constrained to work within the rotational subgroup of the molecular point group. However, he showed (and it follows easily from some results on induced representations) that relative statistical weights of rovibronic levels are given correctly when calculated in any subgroup of the group of all nuclear permutations or permutation-inversions.

The next major advance came with the work of Hougen [89-92], who extended Wilson's classification scheme to include improper point group operations. The induced action upon rotational, vibrational and electronic coordinates of nuclear permutations corresponding to point group operations was given explicitly, and it was proposed that permutations inducing improper rotations of molecular axes should be accompanied by an inversion of the molecule in the centre of mass. Since this ensures that all point group operations induce proper transformations of rotational variables, it is possible to conclude that the problem of applying sense-reversing operations to functions of the rotational coordinates is "avoided rather than solved" in Hougen's approach ([90] p.360). Nevertheless, the scheme can be applied consistently, and molecules with unpaired electrons are easily included within the formalism. In the case of nonplanar molecules, the method provides more information than Wilson's approach, and correspondingly detailed selection rules are obtained for transitions between rovibronic states.

It was left to Longuet-Higgins [1] to make quite explicit the connection of the so-called permutation-inversion (PI) group with the fundamental symmetries of the complete molecular Hamiltonian, and to introduce the key concept of feasible operations as transformations "which can be achieved without passing over an insuperable energy barrier", i.e., within the timescale of experimental observation. The group of transformations for rigid molecules considered by Hougen, which is isomorphic with the molecular point group, therefore consists of all feasible nuclear permutations and permutation-inversions. Longuet-Higgins also described a general theory of the symmetry properties of nonrigid molecules, and we shall return to this problem in Chapter 3.

The discussion by Mills [93] of rovibronic selection rules in symmetric tops according to Hougen's theory should be noted, as should Bunker and Papoušek's extension of the theory to linear molecules [94], and Oka's treatment of the parity of molecular wavefunctions [95].

The above work constitutes the core of the orthodox PI approach to rigid molecule symmetry, and this viewpoint has recently been very clearly summarized by Hougen [92] (in an article dealing in detail with the symmetry properties of the methane molecule) and by Bunker [96,97] ([96] is a comprehensive review of PI theory, while [97] is a textbook on the subject).

We must now consider some works that dissent, to a greater or lesser extent, from the mainstream view of rigid molecule symmetry. Recent activity in this area has led to the formulation of several 'new' approaches, and it is clear that some clarification is called for.

First, we mention the alternative scheme for the symmetry labeling of molecular states proposed by Moret-Bailly [98], which has, for example, led to a long-standing disagreement concerning the naming of the vibration-rotation transition in CH_4 that is near-coincident with the 3.39 μm He-Ne laser line [99, 100]. Although apparently only a question of different conventions being adopted for the symmetry classification of rotational wavefunctions (conventions used by Hougen, Moret-Bailly, and Jahn [101] have been compared in detail by Husson [12]), the controversy has deeper aspects [99, 100]. Moret-Bailly has returned to the problem recently (Appendix 1 of [102]), and has given an abstract account of molecular symmetry from a geometrical point of view. We paraphrase the analysis as follows (cf. also [49,56]): in treating the dynamics of rigid or nonrigid molecules, we necessarily establish, through relations such as 1.41 and 1.108, a 1:1 correspondence between nuclear positions in space (the $\{\underset{\sim}{R}^{\alpha}\}$) and reference positions in the molecular frame (the $\{\underset{\sim}{a}^{\alpha}\}$). The concept of molecular symmetry arises from the fact that there is a certain freedom in making the correspondence. Should we choose to associate the vectors $\{\overline{\underset{\sim}{R}^{\alpha}}\}$ rather than $\{\underset{\sim}{R}^{\alpha}\}$ with the $\{\underset{\sim}{a}^{\alpha}\}$, where $\{\overline{\underset{\sim}{R}^{\alpha}}\}$ differs from $\{\underset{\sim}{R}^{\alpha}\}$ by some permutation of identical nuclei, our description of the system is form-invariant. In particular, we can consider all those transformations $\{\underset{\sim}{R}^{\alpha}\} \rightarrow \{\overline{\underset{\sim}{R}^{\alpha}}\}$ induced by congruences of the reference configuration $\{\underset{\sim}{a}^{\alpha}\}$, corresponding to the various ways in which the equilibrium structure $\{\underset{\sim}{a}^{\alpha}\}$ can be set into the nuclear configuration $\{\underset{\sim}{R}^{\alpha}\}$. We thereby obtain a group of permutations isomorphic with the covering group of the (nonplanar) structure $\{\underset{\sim}{a}^{\alpha}\}$, which is identified as the molecular symmetry group. This way of looking at molecular symmetry is very useful in the interpretation of PI operations (§2.3).

A rigorous and very detailed analysis of the fundamental role played by the Eckart conditions in the theory of molecular symmetry has recently been given by Louck and Galbraith [17]. Amongst other things, this important work emphasizes the fact that the conventional Wigner vibrational symmetry operations [84] generate the invariance group of the Eckart frame, a result that provides much insight into Eckart's remarks [9] on the connection between vibrational symmetry and the choice of molecule-fixed axes. The 'symmetry group of the spherical rotor' $O^{\ell}(3) * O^{f}(3)$ is also introduced in [17]; as shown in Appendix 2, this concept facilitates a very clear distinction between the action of 'lab-fixed' and 'molecule-fixed' rotations upon the rotational coordinates.

In their discussion, Louck and Galbraith suggest that there are subtle difficulties in the application and interpretation of conventional PI theory, and conclude that ([17] p. 104): "the group of feasible (permutation or permutation-inversion] operators has little to do with the molecular motions problem - it is the key concept in implementing the Pauli principle". As discussed below (§2.3), this is a conclusion with which we cannot agree. Nevertheless, we acknowledge the inspiration we have derived from Louck and Galbraith's analysis, and note that our treatment of the symmetry properties of nonrigid molecules (Chapter 3) can be regarded as an extension of their rigid molecule formalism.

Cantrell [103] has given a discussion of the application of molecular symmetry to spectroscopy incorporating some of the ideas of Louck and Galbraith, and Cantrell and Galbraith have considered the calculation of statistical weights for octahedral spherical tops [104].

Following Louck and Galbraith, Hilico, Berger and Loete [14] have presented a treatment of vibration-rotation coupling, selection rules, Raman scattering etc. in spherical top molecules, utilizing the double-tensor formalism [11] together with the group chain (cf. Appendix 2)

$$O^{\ell}(3) \times O^{f}(3) \supset O^{\ell}(3) \times G^{f}$$

for classification of wavefunctions and operators. An important feature of the approach is the use of rotational double-tensors (wavefunctions) of definite parity (definition: behaviour under the inversion $\mathcal{J}: C \to -C$).

Berger [105], taking up the suggestion by Louck and Galbraith that it should be possible to base a new account of molecular symmetry on their work on the Eckart frame, has given "a classification for the energy levels [of polyatomic molecules] derived from this new viewpoint". In this chapter, we argue that this viewpoint in fact differs little from the standard PI approach, except for the crucial problem of the parity of rotational wavefunctions.

Very recently, Harter and Patterson have developed a useful and very pictorial formalism for the symmetry labeling of molecular states, determination of multipole selection rules, and calculation of spin statistical weights, which is apparently based upon the frame transformation theory of Fano and Chang [106]. Although Harter and Patterson's work introduces many new and exciting ideas, such as the cluster analysis for large values of angular momentum, and the use of unitary group tableau techniques in molecular physics [107], we are concerned here with those aspects relevant to the fundamental problems of molecular symmetry. Once again, this approach seems to differ from the PI theory only in the treatment of improper rotations of axes, and is identical with that proposed by Berger.

Finally, we mention that Lathouwers [108] has considered the symmetry properties of diatomic molecules within the generator-coordinate method.

Despite the apparent diversity of treatment indicated above, it is possible to view all these approaches from a common standpoint. Although most points are developed in detail in subsequent sections, our main conclusions can briefly be summarized as follows: The analysis by Louck and Galbraith provides a rigorous formal account of the usual PI theory of molecular symmetry. Their careful discussion of the induced action of point group operations and corresponding permutations or permutation-inversions of identical nuclei is particularly useful. Confusion concerning the PI theory appears to have arisen as a result of Hougen's choice of a group of symmetry operations anti-isomorphic with the usual molecular covering group. Together with the formulation of rotational wavefunctions as polynomials in the elements of the direction-cosine matrix ([7]; cf. Appendix 2), the work of Louck and Galbraith points the way to a correct approach to the parity of molecules through use of discontinuous rotational wavefunctions, i.e., PI theory 'without the inversion'. This lead has been followed both by Berger and by Harter and Patterson. In the latter case the connection with PI theory is fairly clear, whereas in the former the relation with the usual formalism is in our opinion obscured by the assumption that Louck and Galbraith did in fact determine a new invariance group of the molecular Hamiltonian. We show below that all these approaches are very closely related to PI theory itself. The earlier classification scheme proposed by Moret-Bailly also follows the same line (use of "fonctions d'ondes discontinues" [12]), except that an arbitrary choice is made for the parity of the rotational wavefunction associated with the ground vibronic state. Nevertheless, difficulties arising from such arbitrariness (pointed out by Hougen, [92] p.110) disappear when full account is taken of the fact that both parities may occur with a given rovibronic state. To be specific, electric-dipole transitions (for example) will connect states having opposite parities, whereas intramolecular perturbations can only link states with the same parity.

Two major considerations lead us to advocate the use of the Berger-Harter-Patterson modification of PI theory.

First, in the case of linear and planar molecules, where there is a homomorphism from the molecular point group onto the group of feasible nuclear permutations (so that the molecules can 'invert' by means of a proper overall rotation), conventional PI theory does not do justice to the full symmetry of the problem. Thus, rovibronic levels for such molecules actually have well-defined parities, which may be positive or negative, so that levels do not occur in practically-degenerate parity doublets as in nonplanar species. It seems to us that this fact is not sufficiently emphasized in the usual PI account. In particular, the standard PI treatment of the symmetry properties of linear molecules [94], e.g., diatomics, provides symmetry labels whose physical significance is not immediately apparent; furthermore, the suggested departure from the standard symmetry labeling scheme [109] and the corresponding alteration of the straightforward selection rules seems an unnecessary complication (§2.4).

Second, and again on the subject of rotational wavefunctions, we note that there can be difficulties with the notion of molecular 'forms' required in the conventional PI discussion of nonplanar species. A 'form' is a coherent superposition of parity eigenstates having a definite (right or left) handedness

$$\psi_{R,L} = \frac{1}{\sqrt{2}} (\psi_+ \pm \psi_-) ,$$

and feasible PI operations are expressly formulated so as not to interconvert left- and right-handed forms. Since tunneling between forms is extremely slow in nonplanar molecules [110], it is appropriate, according to the usual argument, to consider the symmetry properties of the rovibronic wavefunction associated with a given form [96]. This argument is compelling as far as it goes. However, it can happen that, owing to the requirements of the Pauli principle on the total molecular wavefunction, one of the parity eigenstates required to make the forms <u>does not exist</u>; in that case, the rovibronic state of the <u>nonplanar</u> molecule has a definite parity, since one partner of the parity doublet has zero statistical weight. A particularly apposite example is the ground rovibronic (j = 0) state of methane (CH_4), which must have the symmetry species 0^-A_2 in $0^\ell(3) \times T_d$ [105] (see below), and therefore negative parity. The PI prescription fails here, as the two usual options for the theory are not tenable: it is neither possible to deal with right- and left-handed forms, nor is it possible to assume that all rotational wavefunctions have a particular (positive = even) parity (cf. Hougen [92] p. 98).

For molecules whose covering symmetry groups do not contain improper rotations, each rovibronic level is always associated with a (±) parity pair, and there is then no conflict with the traditional picture of two chiral molecular forms [85] (which are rendered 'robust' by interaction with the electromagnetic field [111]).

The plan for the remainder of this chapter is as follows: In §2.1, we present a detailed but otherwise standard (following Longuet-Higgins [1]) discussion of the fundamental symmetries of the complete molecular Hamiltonian. It is emphasized that our usual notions of molecular symmetry ultimately derive from the presence of several identical nuclei, and that molecules generally exhibit spontaneously-broken permutation-inversion symmetry [31-34, 112] characterized by a clustering of energy levels. There follows in §2.2 a formal account of the symmetry properties of rigid molecules, based largely upon that of Louck and Galbraith [17]. We define the correspondence between point group operations and permutations of nuclei, and consider the induced action of such permutations upon molecular coordinates. In §2.3 more details are given concerning permutations of nuclei and their relation to point group operations, with particular reference to the concept of the invariance group of the Eckart frame. In §2.4 we give a self-contained account of the symmetry properties of diatomics, based upon the formalism of §2.2; although there are of course no new results here, our discussion serves as an alternative to the standard PI treatment [94], and may provide some insight into the novel exposition given in [83].

In order to ease the burden of detail in this chapter, one important topic is dealt with separately in an appendix. In Appendix 2 we consider the relation between orbital angular momentum and rotational wavefunctions, and obtain the transformation properties of rotational wavefunctions under internal and external rotations. This work develops several points touched upon in Chapter 1 and in Sections 2.2 and 2.3, and should be regarded as fairly essential background.

2.1 Fundamental Symmetries of the Molecular Hamiltonian

Let us consider again the non-relativistic Hamiltonian for an isolated molecule consisting of N nuclei and N_ε electrons, expressed in lab-fixed cartesian coordinates:

$$\hat{H} = \hat{T}_N + \hat{T}_E + \hat{V}_{NN} + \hat{V}_{NE} + \hat{V}_{EE} \qquad 2.1$$

in an obvious notation.

The Hamiltonian 2.2 is invariant under coordinate transformations induced by the following fundamental symmetries [1]:

(a) Translation along the time axis.

(a') Time-reversal: reversal of the directions of the spins and momenta of all particles.

(b) Translations of the origin of the lab frame $\{\hat{\underset{\sim}{\ell}}_i\}$.

(c) Proper rotations of the lab frame.

(c') Inversion of the lab-fixed coordinate frame.

(d) Permutations of the position vectors and spins of identical particles.

(Unitary symmetries associated with the nuclear spin spaces are not considered; cf. [113]). These symmetries remain valid within the Breit-Pauli approximation [114], where relativistic and magnetic corrections to electronic motion are taken to second order in the fine-structure constant.

(a) Transformations induced by elements τ of the time translation group $\mathcal{T}$ are simply

$$\hat{\tau} : t \rightarrow t + \tau \; . \qquad 2.2$$

The generator of the time evolution 2.2 is the isolated molecule Hamiltonian $\hat{H}$, and the associated constant of the motion is the molecular energy

(a') Apart from its significance in establishing Kramers degeneracy [115], time-reversal symmetry is generally useful in the following way: if, for example, we wish to construct an effective rotational Hamiltonian as a polynomial in the components of the total molecular angular momentum, then time-reversal invariance together with the requirement of hermiticity implies that all coefficients of odd powers of the angular momentum vanish [116].

(b) In the absence of external fields, the molecular Hamiltonian is invariant under the space-translation group $\mathcal{S}$ where

$$\hat{s}(\xi) : \underset{\sim}{R}^{\alpha} \rightarrow \underset{\sim}{R}^{\alpha} + \underset{\sim}{\xi} \qquad 2.3a$$

$$\underset{\sim}{R}^{\varepsilon} \rightarrow \underset{\sim}{R}^{\varepsilon} + \underset{\sim}{\xi} \ , \qquad 2.3b$$

and $\underset{\sim}{\xi}$ is an arbitrary translation vector. Translational invariance ensures that the motion of the molecular centre of mass, which is a collective coordinate, can be separated exactly from internal motions, the associated constant of motion being the linear momentum of the molecule (denoted $\underset{\sim}{k}$).

(c) Having chosen a particular origin for the lab frame $\{\hat{\underset{\sim}{\ell}}_i\}$, the Hamiltonian is invariant under rotations in the group $SO^{\ell}(3)$

$$\text{for all } \rho \in SO^{\ell}(3), \ \rho : R_i^{\alpha} \rightarrow R(\rho)_{ij} R_j^{\alpha} \qquad 2.4a$$

$$R_i^{\varepsilon} \rightarrow R(\rho)_{ij} R_j^{\varepsilon} \qquad 2.4b$$

where (det $R(\rho)$) = +1. These are 'active' rotations of all particles with respect to the lab frame, induced by the transformation

$$\rho : \hat{\underset{\sim}{\ell}}_i \rightarrow \hat{\underset{\sim}{\ell}}_j R(\rho)_{ji} \qquad 2.5$$

(further details concerning rotations of axes etc. are given in Appendix 1). Thus, $SO^{\ell}(3)$ is the group of 'external' rotations [17,56,96] generated by the lab-fixed components of $\hat{\underset{\sim}{J}}$, the total angular momentum of the molecule (excluding nuclear spin) with respect to the coordinate origin.

It is convenient to take the origin of $\{\hat{\underset{\sim}{\ell}}_i\}$ to be at the molecular centre of mass at all times, so that we step into the $\underset{\sim}{k} = \underset{\sim}{0}$ inertial frame. This is not strictly necessary [99], but "corresponds to the point of view of spectroscopy" [31, 115].

Molecular states can therefore be labeled with two quantum numbers j,m, characterizing transformation properties of the wavefunction under $SO^{\ell}(3)$, i.e.,

$$\forall\ \rho\in O^{\ell}(3), \qquad \rho^{\ell}:\psi^{j}_{m} \rightarrow \sum_{m'} \psi^{j}_{m'}\mathcal{D}^{j}(R(\rho))_{m'm}\ , \tag{2.6}$$

where $\mathcal{D}^{j}$ is the Wigner rotation matrix (Appendix 1).

In the absence of electron spin/orbit coupling (from the Breit-Pauli correction [114]), $\hat{\underset{\sim}{J}}$ refers to the total orbital angular momentum. If the nuclear spin angular momentum $\hat{\underset{\sim}{I}}$ is coupled to $\hat{\underset{\sim}{J}}$ we obtain the grand total angular momentum $\hat{\underset{\sim}{F}}$

$$\hat{\underset{\sim}{F}} = \hat{\underset{\sim}{J}} + \hat{\underset{\sim}{I}} \tag{2.7}$$

with associated quantum numbers F, M_F.

Symmetries (a)-(c) generate the 'geometrical' subgroup of the (connected) Galilei group $\mathcal{G}$ [117], which is the semi-direct product

$$\mathcal{T} \otimes (\mathcal{S} \circledS SO^{\ell}(3)) \subset \mathcal{G}\ . \tag{2.8}$$

(c') The symmetry operation $\mathcal{J}$ inverts all particles in the origin:

$$\mathcal{J}\ :\ R^{\alpha}_{i} \rightarrow -\ R^{\alpha}_{i}\ ,\ R^{\varepsilon}_{i} \rightarrow -\ R^{\varepsilon}_{i}\ . \tag{2.9}$$

In principle therefore, molecular states can be assigned a <u>parity</u> quantum number π. It should be stressed that invariance of the molecular Hamiltonian under inversion does not imply that there is a degeneracy with respect to parity (as asserted by Berger <u>et al</u>: "les niveaux d'énergie sont donc, <u>a priori</u>, dégénérés relativement à la parité" [14]). It is an empirical fact that, for nonplanar molecules with a modest degree of internal excitation, rovibronic levels usually appear in almost-degenerate parity doublets [85, 110].

In the $\underset{\sim}{k} = \underset{\sim}{0}$ inertial frame, the inversion $\mathcal{J}$ does not affect the centre of mass coordinate. If, however, photon recoil effects are taken into account, then in principle it is necessary to consider the parities of translational wavefunctions, as pointed out by Moret-Bailly [99].

Combining symmetries (c) and (c') yields the rotation-inversion group $O^{\ell}(3)$. We have (cf. equation A2.60)

$$\forall\ \rho\in O^{\ell}(3), \qquad \rho\ :\ \psi^{j^{\pi}}_{m} \rightarrow \sum_{m'} \psi^{j^{\pi}}_{m'}(\det R(\rho))^{\nu_{\pi}}\mathcal{D}^{j}(R(\rho))_{m'm} \tag{2.10a}$$

where

$$\nu_\pi \equiv \begin{cases} 0 & \pi = + \ (g) \\ 1 & \pi = - \ (u) \ . \end{cases} \qquad 2.10b$$

The problem of the parity of molecular wavefunctions [95] is concerned with finding rotational wavefunctions spanning IRs of the rotation-inversion group $O^\ell(3)$, as discussed in Appendix 2.

(d) Invariance of the Hamiltonian with respect to all permutations of the variables corresponding to identical particles reflects the essential indistinguishability of identical micro-particles. In this context, the term 'identical particles' refers either to electrons, or to nuclei having identical charges, masses and states of internal excitation.

Let there be N_a identical particles of type a in the molecule. The Hamiltonian is therefore invariant under S_{N_a}, the symmetric group of degree N_a

$$\forall \ \mathcal{P} \in S_{N_a}, \quad \mathcal{P} : \underset{\sim}{R}^\alpha \rightarrow \underset{\sim}{R}^{\bar{\alpha}} \qquad 2.11a$$

$$\mathcal{P} : \sigma^\alpha \rightarrow \sigma^{\bar{\alpha}} \ , \qquad 2.11b$$

where σ^α is the spin variable associated with particle α, and the labels $\{\bar{\alpha}\}$ are related to the set $\{\alpha\}$ by the permutation $\mathcal{P}$.

In principle, molecular states can be classified with respect to the permutation group

$$S_{N_\varepsilon} \times \prod_a S_{N_a} \qquad 2.12$$

where the direct product extends over all sets of identical nuclei. The spin-statistics relation (Pauli principle) is then stated as follows: if the spin of particle type a is integral/half-integral then the IR of S_{N_a} spanned by the total molecular wavefunction must be the symmetric($[N_a]$)/alternating($[1^{N_a}]$) (Pauli) representation.

To summarize, the direct product group

$$SO^\ell(3) \times (E, \mathcal{J}) \times S_{N_\varepsilon} \times \prod_a S_{N_a} \qquad 2.13a$$

or

$$O^\ell(3) \times S_{N_\varepsilon} \times \prod_a S_{N_a} \qquad 2.13b$$

is the fundamental symmetry group of the many-particle molecular Hamiltonian. When there is only a single nucleus present, so that the 'molecular' Hamiltonian 2.1 describes an atom containing N_ε electrons, we obtain the familiar symmetry group

$$O^{\ell}(3) \times S_{N_\varepsilon} , \qquad 2.14$$

which is extensively exploited in the theory of atomic spectra [115].

The phenomenon of molecular symmetry clearly involves the presence of sets of identical nuclei, and molecular symmetry operations must be related to the action of permutations of nuclei from $\prod_a S_{N_a}$.

As was shown in Chapter 1, the Wilson-Howard-Watson Hamiltonian for rigid molecules (1.98) and its natural generalization to nonrigid molecules (1.125) are obtained from the Hamiltonian expressed in lab-fixed coordinates <u>via</u> transformations (1.43 and 1.108) to Born-Oppenheimer coordinates. Since a coordinate transformation cannot change (but only bring to light or mask) the symmetry properties of a system, it must be concluded that the Hamiltonians 1.98 and 1.125 are <u>formally</u> invariant under the fundamental symmetry group 2.13; more particularly, under the <u>complete nuclear permutation-inversion</u> (CNPI) group [96]

$$(E, \mathcal{J}) \times \prod_a S_{N_a} , \qquad 2.15$$

whose order increases very rapidly with the number of identical nuclei in the molecule. Note that the above discussion emphasizes the external, geometric character of the inversion $\mathcal{J}$, to be contrasted with the intrinsic permutational symmetries $\prod S_{N_a}$. If, following Berger, we contend that "...it is important not to mix symmetry operations related to the molecule-fixed frame and those related to the lab-fixed frame" [105], then we should consider only the permutational subgroup of the CNPI group

$$\prod S_{N_a} \subset (E, \mathcal{J}) \times \prod S_{N_a} \qquad 2.16$$

as relevant for 'internal' molecular symmetry.

The full permutational or CNPI symmetry is not, however, apparent from our chosen molecular coordinates. Indeed, as we shall see in §2.3 for rigid molecules and in Chapter 3 for nonrigid molecules, the only permutations with which we can deal straightforwardly are those related in a strictly defined fashion to the symmetry properties of the underlying molecular model [17, 49]. The induced action of an arbitrary permutation of nuclei (element of $\prod S_{N_a}$) upon the molecular coordinates is not easily determined [13].

Of course, the choice of molecular coordinates reflects the empirical fact that molecules usually exhibit spontaneously-broken permutation-inversion symmetry [33, 112], which is manifest in the clustering of molecular energy levels (cf. [83]; also called structural degeneracy [97]). Each cluster of levels corresponds to a set of irreducible representations of the CNPI group, induced from a given irreducible representation Γ_H of a subgroup $\mathbf{H}$ ([125]; cf. Appendix 3)

$$\Gamma_H \uparrow ((E,\mathcal{J}) \times \prod S_{N_a}) \qquad 2.17$$

where

$$\mathbf{H} \subset (E,\mathcal{J}) \times \prod S_{N_a} \qquad 2.18$$

is defined as the molecular symmetry group (MSG).

Precisely what constitutes a quasi-degenerate cluster depends on the available experimental resolution - in other words, the appropriate MSG depends upon the time-scale of interest [61]. It is the job of a useful group-theoretical description to provide a sufficient yet non-redundant labeling of the clusters [118], rather than the individual molecular states, and this task is accomplished (by definition) through the group of feasible transformations.

In the next section, we turn to the particular case of rigid molecules, where the molecular symmetry group $\mathbf{H}$ is isomorphic with or a homomorphic image of a point symmetry group $\mathbf{G}$.

2.2 Symmetry Properties of Rigid Molecules - Formalism

In this section we develop a formal approach to the symmetry properties of rigid molecules. Although much influenced by the analysis of Louck and Galbraith [17], our account also provides a basis for understanding other points of view.

We recall from Chapter 1 that the concept of the static molecular model, represented by the 3N vector components $\{a_i^\alpha\}$, underlies our description of the dynamics of rigid molecules (§1.2). Let us now consider the geometric symmetry of the static model.

A rotation or rotation-inversion acts upon the static model vector $\underset{\sim}{a}^{\alpha}$ as follows:

$$\rho : a_i^{\alpha} \rightarrow R(\rho)_{ij}\, a_j^{\alpha} , \qquad 2.19$$

where $R(\rho)$ is a rotation matrix defined in Appendix 1, so that ρ is an active rotation of the static model. A rotation or rotation-inversion g is said to be a <u>symmetry operation of the static molecular model</u> if it induces a permutation of static model vectors associated with identical nuclei, i.e., if

$$g : a_i^{\alpha} \rightarrow R(g)_{ij} a_j^{\alpha} = \sum_{\beta} a_i^{\beta} S(g)_{\beta\alpha} , \qquad 2.20$$

where $S(g)$ is an N by N permutation matrix, which permutes identical nuclei (and so commutes with the mass matrix). The operation g is a covering symmetry of the static model $\{\underset{\sim}{a}^{\alpha}\}$. The set of all such symmetry operations forms a group G, which is just the <u>point group of the static molecular model</u>

$$G \equiv \{g \mid R(g)_{ij} a_j^{\alpha} = \sum a_i^{\beta} S(g)_{\beta\alpha}\}. \qquad 2.21$$

There are three representations of G that are of particular interest. First, the faithful defining representation by 3 by 3 orthogonal matrices

$$\{R(g) \mid g \in G\}, \quad \text{where} \quad R(g_1)R(g_2) = R(g_1 g_2). \qquad 2.22$$

Second, the representation by N by N permutation matrices, which is not necessarily faithful:

$$\{S(g) \mid g \in G\}, \quad \text{where} \quad S(g_1)S(g_2) = S(g_1 g_2) . \qquad 2.23$$

Third, the faithful representation by permutation-rotations (<u>perrotations</u> [56]) L_g,

$$L_g \equiv R(g) \otimes S(g) \qquad \text{for all } g \in G, \qquad 2.24a$$

with

$$L_{g_1 g_2} = L_{g_1} \cdot L_{g_2} . \qquad 2.24b$$

The realization of the molecular point group by perrotations

$$L_G \equiv \{L_g \mid g \in G\} \qquad 2.25$$

was introduced by Wigner [84] in the study of vibrations, and 2.24 is the usual 3N by 3N external cartesian coordinate representation [6].

Although a simple consequence of 2.20, it is nevertheless important to note that, when acting upon the static model vectors $\{\underset{\sim}{a}^{\alpha}\}$, every perrotation is equivalent to the identity, i.e., for all operations g in G,

$$L_g \cdot a_i^{\alpha} = \sum R(g)_{ij} S(g)_{\alpha\beta} a_j^{\beta} = a_i^{\alpha} \ . \qquad 2.26$$

The point group G can therefore be characterized by the set $\{L_g\}$ of transformations that leave the equilibrium configuration $\{\underset{\sim}{a}^{\alpha}\}$ invariant, a result which is of significance when we come to consider the invariance group of the Eckart frame (§2.3).

Several static model representations have been considered in detail by Louck and Galbraith [17] ($XY_4(T_d)$, $XY_5(D_{3h})$ and $XY_6(O_h)$). In Figure 2.1 we show the XY_3 static model having C_{3v} covering symmetry, while Table 1 lists the associated rotation matrices R(g) and permutation matrices S(g).

Let G^{π} be the group of all distinct permutations of nuclei associated with the point group G through 2.20. G^{π} is clearly a subgroup of $\prod_a S_{N_a}$, the group of all permutations of identical nuclei (2.16)

$$G^{\pi} \subset \prod_a S_{N_a} \ . \qquad 2.27$$

We distinguish three possible cases:

(a) Nonplanar static model. In this case, the group of distinct permutations is isomorphic with the point group of the molecular model

$$G^{\pi} \overset{\text{iso}}{=} G \ . \qquad 2.28$$

(b) Planar static model. Here, there is a 2:1 homomorphism from the point group G onto G^{π}, since reflection in the plane of the molecule is equivalent to the identity permutation of nuclei

$$G \xrightarrow{\text{ho}} G^{\pi} \qquad \text{2:1 homomorphism} \ . \qquad 2.29$$

(c) Linear static model. In this case, there is an ∞:1 homomorphism from G onto G^{π} (whether the point group is $C_{\infty v}$ or $D_{\infty h}$), since all rotations about the molecular axis or reflections in a plane containing the molecular axis are equivalent to the identity (cf. [94]):

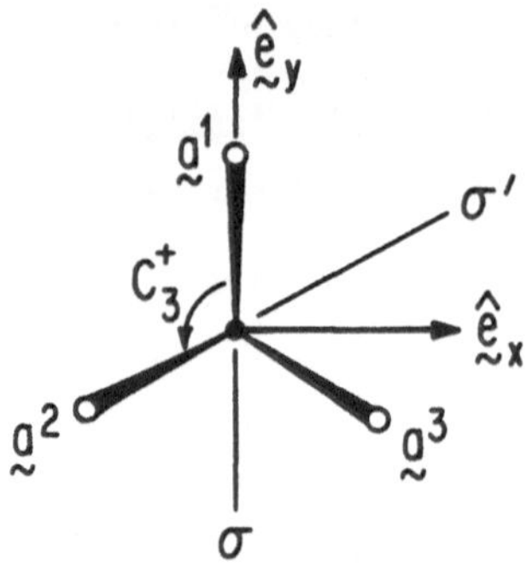

Figure 2.1 The $XY_3 - C_{3v}$ static molecular model

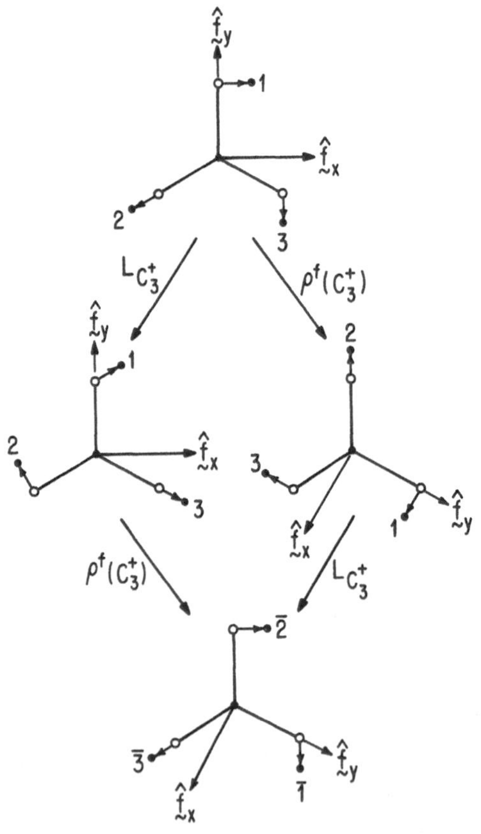

Figure 2.2 The permutation $\wp(C_3^+)$. The lab frame $\{\hat{\ell}_i\}$ maintains an arbitrary but fixed orientation with respect to the page.

Table 1: Rotation matrices R(g) and permutation matrices S(g) associated with elements of the symmetry group C_{3v} of the XY_3 static model shown in Figure 2.1. (Permutations act only on the 3 identical Y particles).

$$G = C_{3v} = \{\hat{E}, \hat{C}_3^+, \hat{C}_3^-, \hat{\sigma}, \hat{\sigma}' = \hat{\sigma}\cdot\hat{C}_3^+, \hat{\sigma}'' = \hat{\sigma}\cdot\hat{C}_3^-\}$$

$$R(E) = \begin{bmatrix} 1 & 0 & 0 \\ 0 & 1 & 0 \\ 0 & 0 & 1 \end{bmatrix} \qquad S(E) = \begin{bmatrix} 1 & 0 & 0 \\ 0 & 1 & 0 \\ 0 & 0 & 1 \end{bmatrix}$$

$$R(C_3^+) = \begin{bmatrix} -1/2 & -\sqrt{3}/2 & 0 \\ \sqrt{3}/2 & -1/2 & 0 \\ 0 & 0 & 1 \end{bmatrix} \qquad S(C_3^+) = \begin{bmatrix} 0 & 0 & 1 \\ 1 & 0 & 0 \\ 0 & 1 & 0 \end{bmatrix}$$

$$R(C_3^-) = \begin{bmatrix} -1/2 & \sqrt{3}/2 & 0 \\ -\sqrt{3}/2 & -1/2 & 0 \\ 0 & 0 & 1 \end{bmatrix} \qquad S(C_3^-) = \begin{bmatrix} 0 & 1 & 0 \\ 0 & 0 & 1 \\ 1 & 0 & 0 \end{bmatrix}$$

$$R(\sigma) = \begin{bmatrix} -1 & 0 & 0 \\ 0 & 1 & 0 \\ 0 & 0 & 1 \end{bmatrix} \qquad S(\sigma) = \begin{bmatrix} 1 & 0 & 0 \\ 0 & 0 & 1 \\ 0 & 1 & 0 \end{bmatrix}$$

$$R(\sigma') = \begin{bmatrix} 1/2 & \sqrt{3}/2 & 0 \\ \sqrt{3}/2 & -1/2 & 0 \\ 0 & 0 & 1 \end{bmatrix} \qquad S(\sigma') = \begin{bmatrix} 0 & 0 & 1 \\ 0 & 1 & 0 \\ 1 & 0 & 0 \end{bmatrix}$$

$$R(\sigma'') = \begin{bmatrix} 1/2 & -\sqrt{3}/2 & 0 \\ -\sqrt{3}/2 & -1/2 & 0 \\ 0 & 0 & 1 \end{bmatrix} \qquad S(\sigma'') = \begin{bmatrix} 0 & 1 & 0 \\ 1 & 0 & 0 \\ 0 & 0 & 1 \end{bmatrix}$$

$$G = C_{\infty v} \xrightarrow{ho} \{E\} \qquad \infty:1 \quad , \tag{2.30a}$$

$$G = D_{\infty h} \xrightarrow{ho} \{E,(12)\} \qquad \infty:1 \quad . \tag{2.30b}$$

The homomorphisms in (b) and (c) have interesting consequences for the symmetry classification of the states of planar and linear molecules, respectively. Thus, for planar molecules, the fact that an improper rotation (reflection in the plane of the molecule) is equivalent to the identity ensures that rovibronic states have well-defined parities [95]. The linear molecule case is discussed in detail in §2.4.

We now come to the main results of this section: each element g of the point group G of the static molecular model has the following action upon the molecular variables:

$$g:C \longrightarrow C\tilde{R}(\rho) \tag{2.31a}$$

or

$$g:C \longrightarrow (\det R(\rho))C\tilde{R}(\rho) \ , \tag{2.31a'}$$

$$g:a_i^\alpha \longrightarrow L_g \cdot a_i^\alpha \equiv \sum_\beta R(g)_{ij} S(g)_{\alpha\beta} a_j^\beta = a_i^\alpha \ , \tag{2.31b}$$

$$g:d_i^\alpha \longrightarrow \overline{d}_i^\alpha \equiv L_g \cdot d_i^\alpha = \sum_\beta R(g)_{ij} S(g)_{\alpha\beta} d_j^\beta \ , \tag{2.31c}$$

$$g:R_i \longrightarrow R_i \ , \tag{2.31d}$$

$$g:r_i^\varepsilon \longrightarrow R(g)_{ij} r_j^\varepsilon \ , \tag{2.31e}$$

The transformations of rotational variables 2.31a are used by Berger [105], Harter and Patterson [83], and Moret-Bailly [12], whereas the feasible transformations 2.31a' (which always preserve the relative sense of the lab- and molecule-fixed frames) are advocated by Hougen [90], Longuet-Higgins [1], Louck and Galbraith [17], and Bunker [96] (and are the conventional choice). For nonplanar rigid molecules, 2.31a is unfeasible when $(\det R(g)) = -1$.

Recalling the relation between molecular variables and the lab-fixed components of nuclear and electronic position vectors (equations 1.2, 1.43):

$$R_i^\alpha - R_i = C_{ij}(a_j^\alpha + d_j^\alpha)$$

$$R_i^\varepsilon - R_i = C_{ij}r_j^\varepsilon ,$$

we find that elements of the point group induce the following (not necessarily distinct) transformations of position vectors

$$g:\underset{\sim}{r}^\alpha : \longrightarrow \overline{\underset{\sim}{r}^\alpha} \equiv \sum S(g)_{\alpha\beta}\underset{\sim}{r}^\beta \qquad 2.32$$

or $$g:\underset{\sim}{r}^\alpha \longrightarrow (\det R(g))\overline{\underset{\sim}{r}^\alpha} , \qquad 2.32'$$

and $$g:\underset{\sim}{r}^\varepsilon \longrightarrow \underset{\sim}{r}^\varepsilon \qquad 2.33$$

or $$g:\underset{\sim}{r}^\varepsilon \longrightarrow (\det R(g))\underset{\sim}{r}^\varepsilon . \qquad 2.33'$$

The unprimed equations 2.32,33 are a consequence of the choice 2.31a for transformations of rotational variables, and show that elements of the point group G induce permutations ($\mathcal{P}_g \in G^\pi$) of the nuclear position vectors $\{\underset{\sim}{r}^\alpha\}$, where the correspondence between point group operations and permutations is defined by the relation 2.20. Use of the transformations 2.31a is therefore associated with the group chain

$$(E,\mathcal{I}) \times \prod_a S_{N_a} \supset \prod_a S_{N_a} \supset G^\pi . \qquad 2.34$$

Alternatively, the primed equations 2.32',33' result from the use of feasible transformations of rotational variables 2.31a', and show that in this case elements of the point group induce either permutations ((det R(g)) = +1) or permutation-inversions ((det R(g)) = -1) where the correspondence between point group operations and permutations is defined exactly as above. The PI group of feasible permutations and permutation-inversions is associated with the group chain

$$(E,\mathcal{I}) \times \prod_a S_{N_a} \supset G^\pi(PI), \qquad 2.34'$$

which is distinct from that in 2.34.

In both cases, however, the overall coordinate transformations induced by elements of the point group G are symmetry operations of the complete molecular Hamiltonian (§2.1). It follows that, regardless of the extent of coupling between various internal degrees of freedom, we have a means of classifying molecular state^

and energy levels with respect to a group isomorphic with the covering group of the notional nuclear equilibrium configuration. For 'rigid' molecules, this provides an optimal labeling scheme for the clusters of levels associated with the spontaneously- broken CNPI symmetry (except that, when adopting the group chain 2.34, we necessarily label individually each partner of a parity doublet; however, if one of the partners has zero statistical weight, then the chain 2.34 is particularly appropriate: see below). For 'nonrigid' molecules, this labeling is by definition insufficiently detailed, and it is necessary to insert a supergroup into the chains

$$\prod S_{N_a} \supset H^{\pi} \ (\supset G^{\pi}) \qquad 2.35$$

or

$$(E, \mathcal{J}) \times \prod S_{N_a} \supset H^{\pi}(PI) \quad (\supset G^{\pi}(PI)) \qquad 2.35'$$

This problem is taken up in Chapter 3.

Use of Born-Oppenheimer coordinates means that we write the spatial molecular wavefunction as a product (or sum of products) of rotational, vibrational and electronic functions

$$\psi_{rve} \sim \psi_{rot}(C) \cdot \psi_{vib}(Q_\lambda) \cdot \psi_{el}(r_i^\varepsilon) \quad ,$$

and we must consider the induced action of the coordinate transformations

$$g:(C,d_i^\alpha,r_i^\varepsilon) \rightarrow (C\tilde{R}(g),\ L_g \cdot d_i^\alpha,\ R(g)_{ij} r_j^\varepsilon) \qquad 2.36$$

or

$$g:(C,d_i^\alpha,r_i^\varepsilon) \rightarrow ((\det R(g)C\tilde{R}(g),\ L_g \cdot d_i^\alpha,\ R(g)_{ij} r_j^\varepsilon) \qquad 2.36'$$

upon the individual factors.

Rotational wavefunctions

It is here, and only here, that the difference between the conventional PI approach (2.36) and that proposed by Berger, Harter and Patterson (2.36') is apparent.

Let us recall from Appendix 2 that rotational wavefunctions for an arbitrary molecule can be written as linear combinations of the rotational double-tensors (A2.52)

$$\langle C|j^{\pi};\ m,k\rangle \qquad k,m = j,j-1,\ldots,-j \quad , \qquad 2.37$$

which are functions of the direction-cosine matrix (rotational coordinates) C spanning an IR of the symmetry group of the spherical rotor (§A2.2)

$$O^{\ell}(3) * O^{f}(3) \quad . \tag{2.38}$$

A typical element of $O^{\ell}(3) * O^{f}(3)$, denoted (ρ^{ℓ},ρ^{f}), induces the transformation of rotational coordinates

$$(\rho^{\ell},\rho^{f}):C \rightarrow R(\rho^{\ell})C\tilde{R}(\rho^{f}) \tag{2.39}$$

and has the following action upon rotational double-tensors (A2.62)

$$(\rho^{\ell},\rho^{f}):|j^{\pi};m,k\rangle \rightarrow \sum_{m'}\sum_{k'} (\det R(\rho^{\ell}))^{\nu_{\pi}}(\det R(\rho^{f}))^{\nu_{\pi}}$$

$$|j^{\pi};m'k'\rangle\mathcal{D}^{j}(R'(\rho^{\ell}))_{m'm}\mathcal{D}^{j}(R'(\rho^{f}))_{k'k} \tag{2.40}$$

where (cf. 2.10) $R(\rho^{\ell}) \equiv (\det R(\rho^{\ell}))\ R'(\rho^{\ell})$ 2.41

and $R(\rho^{f}) \equiv (\det R(\rho^{f}))\ R'(\rho^{f})$. 2.42

The quantum number j is the total molecular (orbital) angular momentum, while m is the projection onto the $\hat{\ell}_z$-axis. The quantum number π is the parity of the rotational wavefunction, and may be either positive or negative. Moreover, π is the parity of the molecule: since inversion in the centre of mass affects only the rotational coordinates

$$\mathcal{J} :(C,\ d_i^{\alpha},\ r_i^{\varepsilon}) \rightarrow (-C,d_i^{\alpha},r_i^{\varepsilon}) \tag{2.43}$$

and since from 2.40

$$\mathcal{J} :|j^{\pi};m,k\rangle = (-1)^{\nu_{\pi}}|j^{\pi};m,k\rangle \quad , \tag{2.44}$$

it is clear that π determines the molecular parity. Hence, use of the rotational wavefunctions $|j^{\pi};m,k\rangle$, which are defined on both disjoint domains of the matrix variable C (det C = ± 1), corresponds to considering molecular states of well-defined parity. The quantum number k is a projection of the total angular momentum onto the molecule-fixed axes.

Now, it is apparent from the above that the transformation of rotational variables induced by the point group operation g is an element of the group of 'molecule-fixed' rotations $O^{f}(3)$, or $SO^{f}(3)$, as the case may be

$$g^f : C \rightarrow C\tilde{R}(g) \qquad g^f \in G^f \tag{2.45}$$

<u>or</u>

$$g^f : C \rightarrow (\det R(g))C\tilde{R}(g) \equiv C\tilde{R}'(g) \qquad g^f \in G^f(PI). \tag{2.45'}$$

The transformation properties of rotational wavefunctions under G are then given unambiguously by

$$g:|j^\pi;m,k\rangle \rightarrow \sum_{k'} |j^\pi;m,k'\rangle\, (\det R(g))^{\overset{\nu}{\pi}} \mathcal{D}^j(R'(g))_{k'k} \tag{2.46}$$

<u>or</u>

$$g:|j^\pi;m,k\rangle \rightarrow \sum_{k'} |j^\pi;m,k'\rangle \mathcal{D}^j(R'(g))_{k'k} \;. \tag{2.46'}$$

It can be seen immediately that it is not quite true that the PI approach (2.46') is restricted to rotational functions having a particular (positive) parity ([92], p.98), but rather that <u>feasible transformations do not distinguish between functions with different parities</u>. The linear combinations of parity eigenstates

$$|j^{R,L};m,k\rangle = \frac{1}{\sqrt{2}}\,(|j^+;m,k\rangle \pm |j^-;m,k\rangle) \tag{2.47}$$

correspond to molecular frameworks with definite handedness, and it follows from 2.46' that feasible permutations and permutation-inversions do not interconvert left- and right-handed forms, whereas when the determinant of the rotation matrix R(g) equals -1 the associated nuclear permutation 2.46 changes the handedness of the molecular framework.

The representations

$$\{(\det R(g))^{\overset{\nu}{\pi}} \mathcal{D}^j(R'(g))\} \equiv j^\pi \qquad \text{in } \; O^f(3) \tag{2.48}$$

<u>or</u>

$$\{\mathcal{D}^j(R'(g))\} \equiv j^+ \qquad \text{in } \; O^f(3) \tag{2.48'}$$

must be reduced into IRs of the group G, and (for example) Herzberg [119] has given tables for the reduction into various common point groups (note also the nomograms introduced in [83]). We write

$$j^\pi \rightarrow \sum_{\Gamma} c_\Gamma(\pi)\Gamma \qquad \text{in } (E,\mathcal{J}) \times G^f \tag{2.49}$$

$$j^+ \rightarrow \sum_{\Gamma'} c_{\Gamma'}\Gamma' \qquad \text{in } G^f(PI) \supset SO^f(3) \tag{2.49'}$$

corresponding to the group chains

$$O^\ell(3) * O^f(3) \supset O(3) * G^f \supset (E,\mathcal{J}) \times G^f \tag{2.50}$$

or $$O^{\ell}(3) * O^{f}(3) \supset SO^{\ell}(3) * G^{f}(PI) \supset G^{f}(PI) \qquad 2.50'$$

respectively.

For the group chain 2.50, the transformations to functions symmetry-adapted to the group $O^{\ell}(3) * G^{f}$ can be written

$$|j^{\pi};\beta\Gamma\gamma\rangle \equiv \sum_{k} |j^{\pi};m,k\rangle\langle j^{\pi}k\,|\beta\Gamma\gamma\rangle \qquad 2.51$$

where γ labels a component of the IR Γ of G, and β is a multiplicity index, needed when $c_{\Gamma} > 1$ [121]; the coefficients of the unitary transformation in equation 2.51 can be determined once and for all for given j^{π} and **G** [172].

Noting that a lab-fixed rotation $\rho^{\ell} \in O^{\ell}(3)$ affects only the rotational coordinates C:

$$\rho^{\ell}:(C,d_{i}^{\alpha},r_{i}^{\varepsilon}) \rightarrow (R(\rho^{\ell})C,\ d_{i}^{\alpha},r_{i}^{\varepsilon}), \qquad 2.52$$

we see that classification of the rotational wavefunctions in the group $O^{\ell}(3)$ is equivalent to classification of the total molecular wavefunction with respect to $O^{\ell}(3)$; in this sense, the total angular momentum of the molecule can be said to be carried by the motion of the Eckart frame [7] (indeed, this is true for any molecule-fixed frame for which the co-rotation condition

$$[\hat{J}_{i},C_{jj'}] = i\varepsilon_{ijk}C_{kj'}$$

is valid [30]).

Vibrational wavefunctions

The vibrational wavefunction $\psi_{vib}(Q_{\lambda})$ is taken to be a function of 3N-6 normal coordinates $\{Q_{\lambda}\}$, introduced in §1.3

$$Q_{\lambda} = \sum_{\alpha} \ell_{\lambda,\alpha j}(m_{\alpha}^{1/2}d_{j}^{\alpha}) \qquad \lambda = 1,\ldots,3N\text{-}6 \qquad 2.53$$

(or linear combinations thereof ≡ symmetry coordinates). The symmetry species of the vibrational coordinates are found in the familiar way by reducing the 3N by 3N representation $L_{G} \equiv \{R(g) \otimes S(g)\}$, and subtracting translations and rotations [6]. Knowing the transformation

$$g: d_{i}^{\alpha} \rightarrow \bar{d}_{i}^{\alpha} = L_{g} \cdot d_{i}^{\alpha} \qquad (2.31c)$$

it is straightforward to induce the action of g upon the Q_λ

$$g:Q_\lambda \to \overline{Q}_\lambda \equiv \sum \Gamma(g)_{\lambda\lambda'}Q_{\lambda'} \qquad 2.54$$

and thereby upon the vibrational wavefunction itself

$$g:\psi_{vib} \to \hat{\Omega}_g\psi_{vib} \qquad 2.55a$$

where $$[\hat{\Omega}_g\psi_{vib}](Q_\lambda) \equiv \psi_{vib}(\Gamma(g)^{-1}_{\lambda\lambda'}Q_{\lambda'}) . \qquad 2.55b$$

Electronic wavefunctions

The induced action of point group operations upon (molecule-fixed [83]) electronic wavefunctions $\psi_{el}(r_i^\varepsilon)$ is straightforward. Since

$$g:r_i^\varepsilon \to R(g)_{ij}r_j^\varepsilon \qquad (2.31c)$$

we have $$g:\psi_{el} \to \hat{\Omega}_g\psi_{el} \qquad 2.56a$$

where $$[\hat{\Omega}_g\psi_{el}](r_i^\varepsilon) \equiv \psi_{el}(\tilde{R}(g)_{ij}r_j^\varepsilon) . \qquad 2.56b$$

This is identical with the usual action of a point symmetry operation upon an electronic function [120].

The vibronic product functions ψ_{ve} can be symmetry adapted to the group G. Let us consider a set of vibronic functions spanning a particular IR Γ',

$$\{\psi_{ve}\} \equiv \{|\Gamma'_{ve}\gamma'\rangle\}. \qquad 2.57$$

Using the Wigner coefficients for the group G [121], the vibronic functions can be coupled with the symmetry adapted rotational functions 2.51 to give rovibronic states with well-defined symmetry in the group $O^\ell(3) \times G$:

$$|j^\pi;m,\Gamma''\gamma''\rangle = \sum_{\gamma\gamma'} |j^\pi;m,\beta\Gamma\gamma\rangle\, |\Gamma'_{ve}\gamma'\rangle\langle\Gamma\gamma\Gamma'\gamma' \mid \beta''\Gamma''\gamma''\rangle \qquad 2.58$$

where β'' is a multiplicity index for the reduction $\Gamma \times \Gamma'$.

The rovibronic states $|j^\pi;m,\Gamma''\gamma''\rangle$ are the 'BOA-constricted' states discussed by Harter and Patterson [83], where the symmetry label [Γ''], which determines the transformation properties of the rovibronic wavefunction under nuclear permutations, defines the 'soul' of the molecule. Here, we define the BOA-constricted states directly via 2.58, and do not consider the rather unphysical notion of a lab-fixed vibronic (as opposed to electronic) function (contrast [83], §III).

To summarize: there are two alternative group chains available for the symmetry classification of rigid molecule states :

(1) Berger, Harter and Patterson, Moret-Bailly

$$O^{\ell}(3) * O^{f}(3) \times (G_{vib} \times G_{el})$$
$$\cup$$
$$O^{\ell}(3) * G^{f} \times (G_{vib} \times G_{el})$$
$$\cup$$
$$O^{\ell}(3) \times G \qquad\qquad O^{\ell}(3) \times \prod S_{N_a}$$
$$\text{iso. or ho.} \searrow \qquad \cup$$
$$O^{\ell}(3) \times G^{\pi}$$

(2) Hougen, Longuet-Higgins, Bunker

$$SO^{\ell}(3) * SO^{f}(3) \times (G_{vib} \times G_{el})$$
$$\cup$$
$$SO^{\ell}(3) * G^{f}(PI) \times (G_{vib} \times G_{el})$$
$$\cup$$
$$SO^{\ell}(3) \times G(PI) \qquad\qquad SO^{\ell}(3) \times (E, \mathcal{J}) \times \prod S_{N_a}$$
$$\text{iso. or ho.} \searrow \qquad \cup$$
$$SO^{\ell}(3) \times G^{\pi}(PI)$$

where G_{vib} and G_{el} are the groups of transformations of vibrational (2.31c) and electronic (2.31e) variables, respectively, and G(PI) is isomorphic with the point group G.

Whenever there is a homomorphism from G (G(PI)) onto G^{π} (G^{π}(PI)), there is a corresponding restriction on the IRs of G (G(PI)): the only IRs of G (G(PI)) that are permitted are those for which the characters of all operations equivalent to the identity permutation are equal. This condition implies that rovibronic states of planar and linear molecules have definite parities.

In the remainder of this section, we briefly discuss two important topics; the formulation of selection rules for multipole transitions and intramolecular perturbations, and the calculation of nuclear spin statistical weights, using the group $O^{\ell}(3) \times G$.

First, consider a molecular multipole operator $T^{j^\pi}(\underset{\sim}{r}^\alpha;\underset{\sim}{r}^\varepsilon)$ having tensorial character j^π defined in the lab frame. Such operators are introduced to describe the interaction of a molecular system with an external electromagnetic field, and an important example is the electric-dipole operator ($j^\pi = 1^-$)

$$T^{1^-}(\underset{\sim}{r}^\alpha;\underset{\sim}{r}^\varepsilon) \equiv e[\sum_\alpha z_\alpha \underset{\sim}{r}^\alpha - \sum_\varepsilon \underset{\sim}{r}^\varepsilon]. \qquad 2.59$$

The multipole operators are invariant under all permutations of identical particles, and so transform as

$$j^\pi \cdot A_1 \qquad 2.60$$

in $O^\ell(3) \times G$, where A_1 is the trivial IR of G.

Multipole selection rules are therefore easily stated for the group $O^\ell(3) \times G$ [83, 105]: the matrix element

$$\langle j_1^{\pi_1};m_1,\Gamma_1\gamma_1 \mid T_m^{j^\pi} \mid j_2^{\pi_2};m_2,\Gamma_2\gamma_2\rangle \qquad 2.61$$

necessarily vanishes unless

(i) $\Gamma_1 \equiv \Gamma_2$

(ii) $\pi_1 \cdot \pi_2 = \pi$

(iii) $|j_1 - j_2| \leq j$

(iv) $m_2 - m_1 = m$

Selection rule (i) states that multipole transitions can only connect levels having the same symmetry in G, while (ii) ensures that parity is conserved. (iii) and (iv) are overall angular momentum selection rules (no resolvable hyperfine splitting).

Approximate but more detailed (rotation, vibration and electronic) selection rules can be obtained by projecting the tensor T^{j^π} onto the molecule-fixed axes (cf. [14]).

We proceed similarly for an intramolecular perturbation $\hat{H}'(\underset{\sim}{r}^\alpha;\underset{\sim}{r}^\varepsilon)$, which is invariant under both external rotations and permutations of identical particles, and so spans

$$0^+ \cdot A_1 \qquad 2.62$$

in $O^\ell(3) \times G$. Corresponding selection rules are

(i) $\Gamma_1 \equiv \Gamma_2$

(ii) $\pi_1 = \pi_2$

(iii) $j_1 = j_2$

(iv) $m_1 = m_2$

This simple and straightforward treatment of selection rules follows those of Berger [105] and Harter and Patterson [83]. As is well known [96], the PI theory gives rise to different selection rules (referred of course to IRs of a different group), since the transformations of rotational variables 2.32a' entangle the external inversion $\mathcal{J}$ with the intrinsic permutation symmetries. Hence, for operators of odd parity, the selection rule $\Gamma_1 \equiv \Gamma_2$ on the IRs of G(PI) does not hold [122].

Up to this point we have not considered the nuclear spin wavefunction ψ_{ns}. However, the Pauli principle stipulates that the total molecular wavefunction must transform in a particular fashion under permutations of identical nuclei (spanning the so-called Pauli representation, cf. §2.1), so that the behaviour of ψ_{ns} is of interest here.

In the absence of hyperfine coupling, the total molecular wavefunction is a simple product of ψ_{ns} and the rovibronic function ψ_{rve}:

$$\psi_{tot} = \psi_{ns} \cdot \psi_{rve} \qquad 2.63$$

where we assume that ψ_{rve} spans a particular IR ${}^{rve}\Gamma$ of the point group **G**. Let ${}^{ns}\Gamma$ be the (in general) reducible representation spanned by the nuclear spin functions, i.e.,

$$g:\psi_{ns} \to \hat{\Omega}_g\psi_{ns} \quad , \quad \text{where} \quad [\hat{\Omega}_g\psi_{ns}](\underset{\sim}{R}^{\alpha}) \equiv \psi_{ns}(\tilde{S}(g)_{\alpha\beta}\underset{\sim}{R}^{\beta}), \qquad 2.64$$

and consider the reduction

$$ {}^{ns}\Gamma \otimes {}^{rve}\Gamma = \sum_i c_i {}^{i}\Gamma \qquad 2.65$$

where the sum is over the IRs $\{{}^{i}\Gamma\}$ of **G**. Then, the multiplicity c_p of the Pauli representation ${}^{P}\Gamma$ is the nuclear spin statistical weight of the state (we must, however, remember to take possible parity doubling into account as well).

As an illustration [105], consider the tetrahedral molecule XY_4, where X is a boson, spin I = 0, and the Y's are fermions, spin I = 1/2, e.g., CH_4.

The point group T_d is isomorphic with the group of all 24 permutations of the Y nuclei, so that the Pauli representation is A_2 in T_d, and $(\pm)A_2$ in $(E, \mathcal{J}) \times T_d$. The nuclear spin functions for the molecule span a $2^4 = 16$-dimensional representation, which reduces to

$$^{ns}\Gamma = 5A_1 + E + 3F_2$$

in T_d. It follows that IRs $^{rve}\Gamma$ with nonzero statistical weights are A_2 ($c_{A_2} = 5$), E ($c_E = 1$) and F_1 ($c_{F_1} = 3$). Noting that changing the parity of the rotational wavefunction affects the IR of T_d spanned by ψ_{rve} (specifically, $(+) \longleftrightarrow (-)$ interchanges the subscripts 1 and 2) we can construct the table ([105], Table 1)

Parity Doublet		Total Statistical Weight
$\underline{(+)A_1}$	$(-)A_2$	5
$(+)A_2$	$\underline{(-)A_2}$	5
$(+)E$	$(-)E$	1 + 1 = 2
$(+)F_1$	$\underline{(-)F_2}$	3
$\underline{(+)F_2}$	$(-)F_1$	3

Rovibronic states underlined have $\underline{\text{zero}}$ statistical weight, which shows that for the CH_4 molecule one partner is missing from every parity doublet except for the E states. As stated earlier, this renders the notion of molecular 'form' inapplicable here, and leads us to advocate use of group chain (1) in preference to the PI group chain (2) (cf. also [190]).

2.3 The Symmetry Properties of Rigid Molecules - Interpretation

We now consider several points arising from the interpretation of the formalism presented in the previous section.

(1) The invariance group of the Eckart frame

As pointed out by Louck and Galbraith [17], the operators L_g (2.31b,c) leave the orientation of the Eckart frame unchanged, so that the group of perrotations

$L_G \equiv \{L_g | g \in G\}$ is the invariance group of the Eckart frame $\{\hat{f}_i\}$. This important result had always been implicitly assumed in previous accounts of molecular symmetry (see, for example, [94] p.422).

To prove this, we start with an arbitrary configuration of nuclei $\{r^\alpha\}$, and suppose that we have calculated the molecule-fixed frame $\{\hat{f}_i\}$ using Eckart's procedure (§1.2) (the nuclear displacements $\{d^\alpha\}$ will be assumed to be 'small' in some sense). We can therefore write

$$r^\alpha = \hat{f}_i(a_i^\alpha + d_i^\alpha). \tag{2.66}$$

The operations

$$L_g : a_i^\alpha \to a_i^\alpha \tag{2.31b}$$

and

$$L_g : d_i^\alpha \to \bar{d}_i^\alpha \equiv \sum R(g)_{ij} S(g)_{\alpha\beta} d_j^\beta \tag{2.31c}$$

are then regarded as defining a new configuration of nuclei $\{\bar{r}^\alpha\}$, with

$$\bar{r}^\alpha \equiv \hat{f}_i(a_i^\alpha + \bar{d}_i^\alpha) \ , \tag{2.67}$$

and we seek to show that, using Eckart's procedure to calculate a new molecule-fixed frame $\{\hat{f}_i'\}$ it turns out that $\hat{f}_i' = \hat{f}_i$, i.e., we obtain the same Eckart frame for the new configuration $\{\bar{r}^\alpha\}$.

Recall that

$$\hat{f}_i = F_j(\Gamma^{-1/2})_{ji}, \tag{1.58}$$

where the Eckart vector is

$$F_i \equiv \sum m_\alpha r^\alpha a_i^\alpha \tag{1.56}$$

and the Gram matrix is

$$\Gamma_{ij} \equiv F_i \cdot F_j \ . \tag{1.57}$$

The perrotation L_g transforms the Eckart vector F_i to

$$L_g : F_i \to \bar{F}_i = \sum_\alpha m_\alpha a_i^\alpha \bar{r}^\alpha = \sum_{\alpha\beta} m_\alpha a_i^\alpha \hat{f}_j R(g)_{jj'} S(g)_{\alpha\beta} \hat{f}_{j'} \cdot r^\beta$$

$$= \sum m_\beta R(g)_{ii'} a_{i'}^\beta \ \hat{f}_j R(g)_{jj'} \ \hat{f}_{j'} \cdot r^\beta$$

using 2.20 and the fact that S(g) commutes with the mass matrix, so that

$$\bar{F}_i = \hat{f}_j R(g)_{jj'} \hat{f}_{j'} \cdot F_{i'} \tilde{R}(g)_{i'i} \ . \tag{2.68}$$

Therefore the Gram matrix transforms to

$$\Gamma_{ij} \to \bar{\underset{\sim}{F}}_i \cdot \bar{\underset{\sim}{F}}_j = R(g)_{ii'}\Gamma_{i'j'}\tilde{R}(g)_{j'j} \ , \tag{2.69}$$

which is a similarity transformation by R(g), and

$$(\Gamma^{-1/2})_{ij} \to R(g)_{ii'}(\Gamma^{-1/2})_{i'j'}\tilde{R}(g)_{j'j} \quad . \tag{2.70}$$

It follows that the perrotations L_g leave the Eckart frame invariant:

$$L_g : \hat{\underset{\sim}{f}}_i \to \hat{\underset{\sim}{f}}_i \equiv \bar{\underset{\sim}{F}}_j R(g)_{jj'}(\Gamma^{-1/2})_{j'i'}\tilde{R}(g)_{i'i} = \hat{\underset{\sim}{f}}_i \ , \tag{2.71}$$

as required.

The implications of this result are of some interest. The invariance 2.71 means that we can write

$$L_g:(C;\ d_i^\alpha) \to (C;\ \bar{d}_i^\alpha \equiv L_g \cdot d_i^\alpha) \ , \tag{2.72}$$

showing explicitly that the operation L_g has no effect upon the rotational coordinates C. When using the Eckart constraints, it is therefore possible to consider the set of vibrational symmetry operations $L_{\mathbf{G}}$ entirely independently of any transformations of rotational variables. This well-defined separation of vibrational and rotational symmetry operations shows the intimate connection between use of the Wigner operations (L_g) and a particular choice of molecule-fixed frame (Eckart frame) [9]. Operations of the form L_g do not necessarily leave C invariant when we use non-linear (e.g. principal axis) constraints to define the molecule-fixed frame.

In the approach we have taken, the invariance group $L_{\mathbf{G}}$ of the Eckart frame arises as a natural consequence of an analysis of the induced action of certain permutations of nuclei upon the molecular coordinates. However, Louck and Galbraith [17] stress that $L_{\mathbf{G}}$ is of significance "quite on its own because an invariance group of the Eckart frame is a realization of the point group by a set of transformations between sets of displacements which are compatible with the Eckart frame" ([17], p.80). As should be clear from our discussion here, this seems to be a question of emphasis rather than fundamental principle; nevertheless Louck and Galbraith's viewpoint leads to some useful insights. Hence, it is possible to regard the transformations

$$L_g : \underset{\sim}{d}^\alpha \to \overline{\underset{\sim}{d}}^\alpha \quad ; \quad \underset{\sim}{r}^\alpha \to \overline{\underset{\sim}{r}}^\alpha$$

as resulting from constrained nuclear motion with respect to the stationary Eckart frame; such motions connect nuclear configurations which are isoenergetic in the absence of rotation, and this interpretation immediately brings to mind the concept of isodynamic operation introduced by Altmann [123-125]. In fact, we shall define the group of transformations L_G to be the isodynamic group associated with the molecular point group G, where L_G is isomorphic with G

$$L_G \equiv \{L_g \mid g \in G\} \overset{iso}{=} G \quad . \tag{2.73}$$

Although in the rigid molecule case the isodynamic group is nothing more than the group of Wigner vibrational symmetry operations, we show in the next chapter that the isodynamic group introduced here has a natural generalization to nonrigid molecules.

Examples of isodynamic transformations for rigid molecules are shown in Figure 2.2 (the operation $L_{C_3^+}$) and Figure 2.3 (the operations $L_{C_3^+}$, L_σ and $L_{\sigma'}$).

We note also that the transformations L_g are isometric in the sense of Günthard et al. [48] (cf. Chapter 3).

The invariance property 2.71 comes as no surprise in the light of the least-squares characterization of the Eckart frame [39]. As observed in §1.4, the quantity

$$\mathcal{E} \equiv \sum m_\alpha \underset{\sim}{d}^\alpha \cdot \underset{\sim}{d}^\alpha \tag{1.115}$$

is invariant under any transformation

$$d_i^\alpha \to \sum S_{\alpha\beta} R_{ij} d_j^\beta \tag{1.116}$$

and therefore under any $L_g \in L_G$. Thus, starting with the nuclear configuration $\{\underset{\sim}{r}^\alpha\}$ and set of displacements $\{\underset{\sim}{d}^\alpha\}$, the new configuration $\{\overline{\underset{\sim}{r}}^\alpha\}$ necessarily corresponds to the the same Eckart frame with a new set of displacements $\{\overline{d}_i^\alpha\}$.

(2) Permutations of nuclei

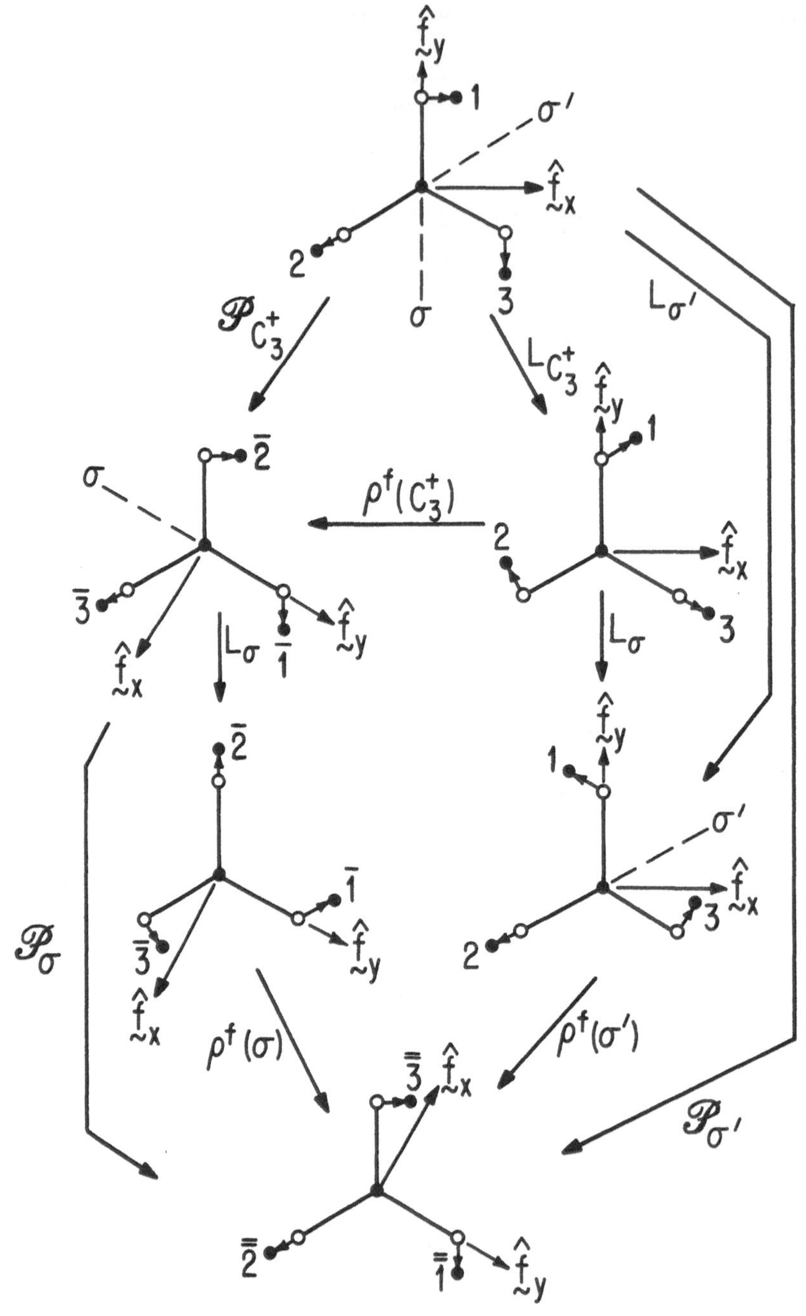

Figure 2.3 The product of permutations $\mathscr{P}(\sigma) \cdot \mathscr{P}(C_3^+) = \mathscr{P}(\sigma')$. The lab frame $\{\hat{\ell}_i\}$ maintains an arbitrary but fixed orientation with respect to the page.

We recall that the transformation of rotational variables (2.31a) associated with the point group operation g is an element g^f of the group $O^f(3)$ of internal rotations,and is defined with respect to the Eckart frame, so that the proper rotation

$$g^f \equiv g^f(\theta, \underset{\sim}{n}^f \equiv \hat{\underset{\sim}{f}}_i n_i^f)$$

corresponds to the following operation (viewed from the molecule-fixed frame): rotate the lab frame $\{\hat{\underset{\sim}{\ell}}_i\}$ together with all lab-fixed objects (i.e., the rest of the universe) through an angle θ about the axis $\hat{\underset{\sim}{n}}^f$ (cf. Appendix 2).

The product of the isodynamic operation L_g and the rotation g^f is a <u>permutation</u> $\mathcal{P}_g$ of the nuclear position vectors ($\mathcal{L}_g$ in the notation of [17], equation 6.16; the electron coordinates are ignored for the moment)

$$\begin{aligned} g^f \cdot L_g(C;d_i^\alpha) &= (C\tilde{R}(g);\ L_g \cdot d_i^\alpha) \\ &= L_g \cdot g^f(C;d_i^\alpha) = \mathcal{P}_g(C;d_i^\alpha) \end{aligned} \qquad 2.74$$

where (2.32)

$$\mathcal{P}_g : \underset{\sim}{r}^\alpha \rightarrow \underset{\sim}{\bar{r}}^\alpha \equiv \sum_\beta S(g)_{\alpha\beta} \underset{\sim}{r}^\beta \quad . \qquad 2.75$$

The diagram

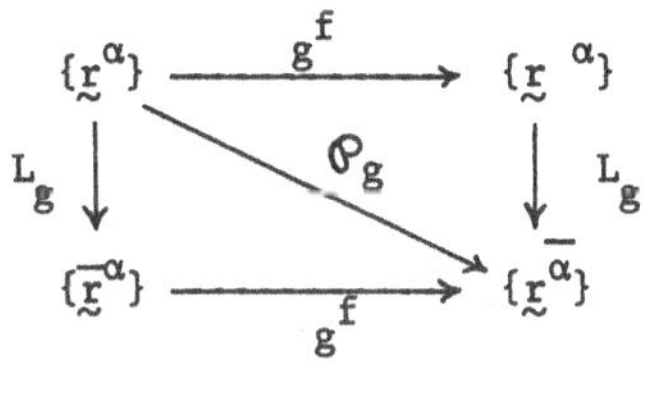

shows that g^f and L_g commute.

These ideas are illustrated in Figure 2.2, which shows the action of the permutation corresponding to the covering operation $\hat{C}_3^+$ of the XY_3 static model (Figure 2.1). We have

$$\hat{C}_3^+ : [\underset{\sim}{a}^1 \underset{\sim}{a}^2 \underset{\sim}{a}^3] \rightarrow [\underset{\sim}{a}^2 \underset{\sim}{a}^3 \underset{\sim}{a}^1] \quad , \qquad 2.76a$$

$$S(C_3^+) = \begin{bmatrix} 0 & 0 & 1 \\ 1 & 0 & 0 \\ 0 & 1 & 0 \end{bmatrix} \qquad \text{(cf. Table 2.1).} \qquad 2.76b$$

A permutation $\mathcal{P}_g$ is described as follows: after the permutation $\mathcal{P}_g$ specified by 2.75, the static model vector $\underset{\sim}{a}^{\alpha}$, which is associated with position vector $\underset{\sim}{r}^{\alpha}$ immediately before the permutation, is associated with the nuclear position vector $\underset{\sim}{r}^{\bar{\alpha}}$,

where $$\underset{\sim}{r}^{\bar{\alpha}} \equiv \sum_{\beta} S(g)_{\alpha\beta}\underset{\sim}{r}^{\beta} .$$

For example, from Figure 2.2 we see that following the permutation $\mathcal{P}(C_3^+)$ the static model vector (equilibrium position) $\underset{\sim}{a}^1$ is associated with position vector $\underset{\sim}{r}^3$, $\underset{\sim}{a}^2$ with $\underset{\sim}{r}^1$, and $\underset{\sim}{a}^3$ with $\underset{\sim}{r}^2$ (cf. 2.76b).

This specification of permutations, based upon the 1:1 correspondence between the static model vectors $\{\underset{\sim}{a}^{\alpha}\}$ and the nuclear positions in space $\{\underset{\sim}{r}^{\alpha}\}$, is very close to the ideas of Moret-Bailly [102] and Gilles and Philippot [56]. It is necessary if we wish to obtain an isomorphic, as opposed to anti-isomorphic, representation of the point group by permutations, as we now show.

Consider the multiplication of two permutations

$$\begin{aligned} \mathcal{P}_{g_2} \cdot \mathcal{P}_{g_1} &= g_2^f \, L_{g_2} \cdot g_1^f \, L_{g_1} = g_2^f g_1^f \cdot L_{g_2} L_{g_1} \\ &= g_{21}^f \cdot L_{g_{21}} = \mathcal{P}_{g_{21}} , \end{aligned} \qquad 2.77$$

where $$g_2 g_1 \equiv g_{21} \in G \qquad 2.78$$

and we note that the crucial commutation rule

$$g_1^f \, L_{g_2} = L_{g_2} \, g_1^f$$

follows from the properties of the isodynamic (Wigner) operations.

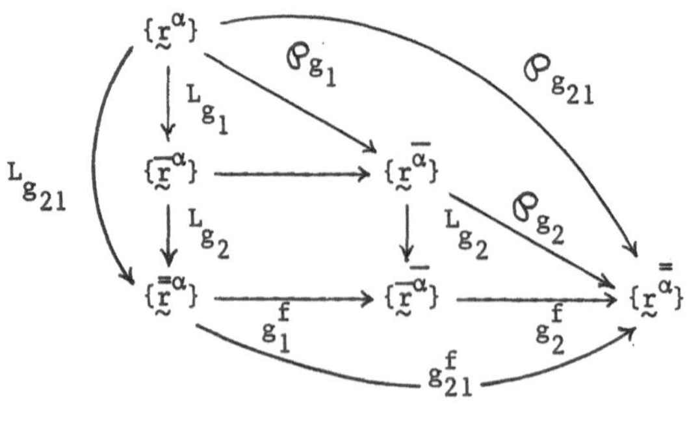

The product rule is illustrated in Figure 2.3, which shows that

$$\wp(\sigma) \cdot \wp(C_3^+) = \wp(\sigma') \qquad 2.79a$$

when $$\hat{\sigma} \cdot \hat{C}_3^+ = \hat{\sigma}' \qquad 2.79b$$

in the point group (Figure 2.1, Table 2.1).

Interpretation of permutations as above therefore provides an isomorphic representation of the molecular point group **G** by the set of permutations

$$\{\wp_g \equiv g^f\ L_g \mid g \in G\} \overset{iso}{=} G.$$

This group appears in all the approaches to molecular symmetry discussed in the introduction, and in particular underlies both the standard PI theory and the formalism of [17] (in both cases only the feasible proper part of g^f being considered). We would contend that there is no essential difference between the usual PI theory and that of Louck and Galbraith, contrary to the assertion ([17] p.104]) quoted in the introduction.

Nonetheless, it is a curious feature of Hougen's account of PI theory [92] that an anti-isomorphic realization of the point group is chosen, and this has led to some confusion. Thus, Louck and Galbraith write a permutation (ostensibly following Hougen) as ([17] equation 6.15a; contrast 2.74)

$$\wp^H_g \equiv g' \cdot L_{g^{-1}} \qquad 2.80$$

where the rotation g' associated with $\wp^H_g$ is held to be an element of $O^\ell(3)$ ([17] p.104), dependent upon the initial value of C

$$g' : C \rightarrow R(g')C \equiv (CR(g)\tilde{C})C \qquad 2.81$$

([17], equations 3.59, 6.15a). From equation A2.49, we see that

$$R(g') = CR(g)\tilde{C} = R(\rho^\ell_{g^{-1}}(C)) \quad , \qquad 2.82$$

i.e., g' is the element of $O^\ell(3)$ associated with the molecule-fixed rotation $(g^{-1})^f$ through the natural anti-isomorphism between $O^\ell(3)$ and $O^f(3)$. The permutations $\wp^H_g$ defined in 2.80 then multiply in anti-isomorphic fashion.

However, this formulation of permutations does not follow from Hougen's original work on the PI group. Thus, according to Hougen ([90], equation (8)) rotations associated with permutations

(i) Commute with lab-fixed components of the total angular momentum, which generate the group $O^{\ell}(3)$, and

(ii) Do not commute with the molecule-fixed components of the total angular momentum, which generate the group $O^{f}(3)$.

In our notation, we have (cf. Appendix 2)

$$[g',\hat{J}_i] = 0 \qquad \text{for all } g',i \quad , \tag{2.83a}$$

and

$$[g',\hat{K}_i] \neq 0 \ . \tag{2.83b}$$

It follows that g' is a molecule-fixed rotation: in fact,

$$g' = (g^f)^{-1} \equiv \bar{g}^f \in O^f(3), \tag{2.84a}$$

i.e.,

$$g' : C \rightarrow C\tilde{R}(\bar{g}) = CR(g) \tag{2.84b}$$

so that the induced action of g' affects the internal (k) rather than the external (m) index of the rotational double-tensor $|j;m,k\rangle$, as required ([90] equation (9)).

The permutations $\wp^H$ then form an <u>anti</u>-isomorphic representation of the point group, and this corresponds exactly to the conventions adopted by Hougen [92] (overbars indicate inverse operations):

$$\begin{aligned} \wp^H_{g_2} \cdot \wp^H_{g_1} &= \bar{g}_2^f L_{\bar{g}_2} \cdot \bar{g}_1^f L_{\bar{g}_1} \\ &= \bar{g}_2^f \bar{g}_1^f \cdot L_{\bar{g}_2 \bar{g}_1} = \overline{g_1 g_2}^f \cdot L_{\overline{g_1 g_2}} = \wp^H_{g_1 g_2} \ . \end{aligned} \tag{2.85}$$

Hence, the only difference between the formalism of [17] and the Hougen/Longuet-Higgins [1,89] PI theory seems to be whether we have an isomorphic [17] or anti-isomorphic [89] representation of the point group by permutations. An isomorphism is clearly to be preferred.

Furthermore, it is easily shown that, using Hougen's procedure for inducing the action of a point group operation upon the molecular wavefunction and <u>retaining</u>

Hougen's labels for symmetry operations [89, 90, 92] (which, when traced back to their definition in terms of their action upon the static model, are seen to correspond to the inverse of 'active' point group operations as usually defined), we obtain exactly the same result as that obtained upon application of the coordinate transformation 2.74 according to Appendix 1, equation A1.11.

In conclusion, the main point is that there are two anti-isomorphisms arising in the molecular symmetry problem, which must be carefully distinguished if we are not to run into confusion:

(a) The natural anti-isomorphism between $O^{\ell}(3)$ and $O^{f}(3)$ (Appendix 2), and

(b) Hougen's anti-isomorphic realization of the point group.

2.4. Symmetry Properties of Diatomic Molecules

In this section we give a short yet self-contained account of the symmetry properties of diatomic and linear molecules, based upon the ideas of the previous sections. Our straightforward derivation of the familiar, physically significant symmetry labels [109] (cf. also [83]) should be compared with the PI treatment by Bunker and Papousek [94].

(1) Heteronuclear Diatomics

For the moment we consider a heteronuclear system, for which the static molecular model has $C_{\infty v}$ covering symmetry (Figure 2.4).

Here, all operations in the point group $C_{\infty v}$ are equivalent to the identity permutation of static model vectors:

$$\hat{C}_z(\theta) : \underset{\sim}{a}^{\alpha} \to \underset{\sim}{a}^{\alpha} \qquad \alpha = 1,2 \tag{2.86a}$$

$$\hat{\sigma}(yz) : \underset{\sim}{a}^{\alpha} \to \underset{\sim}{a}^{\alpha} \tag{2.86b}$$

$$\hat{C}_z(\theta)\, \hat{\sigma}\, \hat{C}_z^{-1}(\theta) : \underset{\sim}{a}^{\alpha} \to \underset{\sim}{a}^{\alpha} \quad , \tag{2.86c}$$

so that we have an ∞:1 homomorphism of $C_{\infty v}(\mathbf{G})$ onto the trivial group of nuclear permutations consisting of the identity ($\mathbf{G}^{\pi}$).

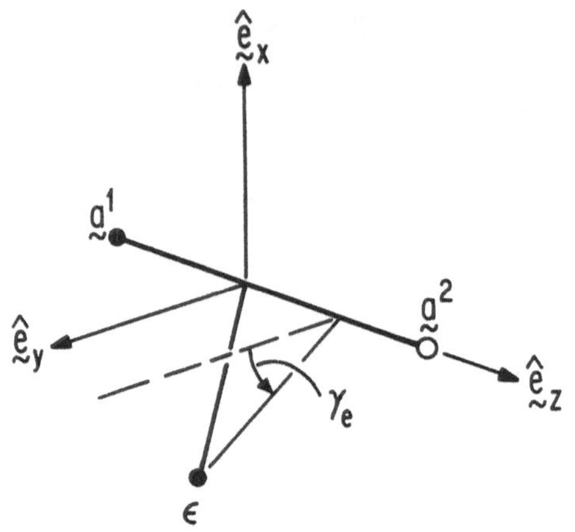

Figure 2.4 The Heteronuclear $C_{\infty v}$ diatomic molecule model.

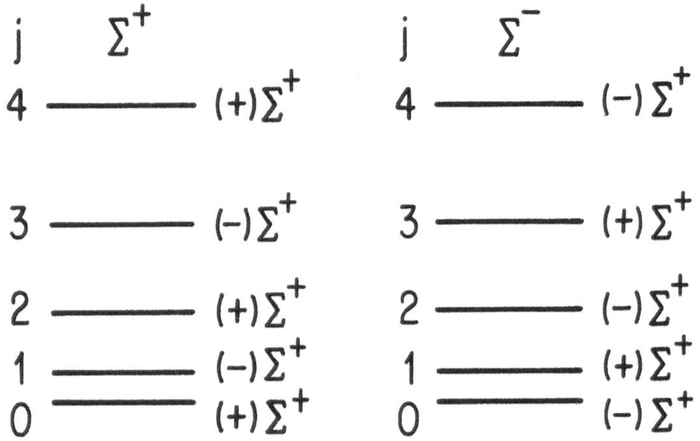

Figure 2.5 Symmetry labels for rotational levels of a heteronuclear diatomic. Σ^+ and Σ^- vibronic states.

The symmetry of the molecular rovibronic function is determined by the product of the symmetries of the rotational wavefuntions $|j^{\pi};m,k\rangle$ and the electronic function $|\Lambda_e\rangle$, since in diatomics the vibrational coordinate (internuclear distance) is totally symmetric.

The orientation of the molecule-fixed frame $\{\hat{\underset{\sim}{f}}_i\}$ is described by an orthogonal matrix C, where $\hat{\underset{\sim}{f}}_z$ always points along the internuclear axis, and the arbitrariness in the disposition of the $\hat{\underset{\sim}{f}}_{x,y}$ axes is reflected in the constraint 2.106 below.

From §2.2, we have the induced action of any operation g in $C_{\infty v}$,

$$g:C \rightarrow C\tilde{R}(g) \qquad 2.87$$

and $$g:|j^{\pi};m,k\rangle \rightarrow \sum_{k'} |j^{\pi};m,k'\rangle(\det R(g))^{\nu_{\pi}}\mathcal{D}^{j}(R'(g))_{k'k} \qquad 2.88$$

In particular,

$$\hat{C}^{f}(\theta,\hat{\underset{\sim}{f}}_z) : |j^{\pi};m,k\rangle \rightarrow |j^{\pi};m,k\rangle\, e^{-ik\theta} \qquad 2.89a$$

and $$\hat{\sigma}^{f}(yz) : |j^{\pi};m,k\rangle \rightarrow |j^{\pi};m,-k\rangle(-)^{\nu_{\pi}+j} \qquad 2.89b$$

so that the set of 2j+1 rotational functions $\{|j^{\pi};m,k\rangle;\ k = j,j-1,\dots,-j\}$ reduces in $O^f(3) \supset C^f_{\infty v}$ as follows [119]

$$j^{\pi} \rightarrow \Sigma^{(-)^{\nu_{\pi}+j}} + \Pi + \Delta + \cdots E_{|j|} \qquad 2.90a$$

where $$\Pi \equiv E_1\ ,\ \Delta \equiv E_2\ ,\ \dots\ . \qquad 2.90b$$

In the PI approach, C SO(3) and is described by the 3 Euler angles (α,β,γ) (Appendix 1); the induced action of point group operations is [94]

$$\hat{C}^{f}(\theta,\hat{\underset{\sim}{f}}_z):(\alpha,\beta,\gamma) \rightarrow (\alpha,\beta,\gamma-\theta) \qquad 2.91a$$

and $$\hat{\sigma}^{f}(yz) : (\alpha,\beta,\gamma) \rightarrow (\pi+\alpha,\ \pi+\beta,\ -\gamma) \qquad 2.91b$$

where $$\hat{\sigma}^{f}(yz) = \mathcal{J}\cdot \hat{C}^{f}(\pi,\hat{\underset{\sim}{f}}_x)\ . \qquad 2.91c$$

The position of an electron with respect to the molecule-fixed frame is specified as shown in Figure 2.4; the only coordinate of interest here is the azimuthal angle

γ_e. Following [94], we measure γ_e from the y-axis: this corresponds to the definition of the third Euler angle γ (Figure A1.3), which measures the orientation of the line of nodes with respect to the $\hat{\underset{\sim}{f}}_y$-axis.

A one-electron wavefunction is denoted $|\Lambda_e\rangle$, and is taken to be an eigenstate of $\hat{L}_z$, the component of electronic angular momentum along $\hat{\underset{\sim}{f}}_z$

$$\hat{L}_z|\Lambda_e\rangle = -i\,\partial/\partial\gamma_e|\Lambda_e\rangle = \Lambda_e|\Lambda_e\rangle \qquad 2.92a$$

i.e.,

$$|\Lambda_e\rangle \sim e^{i\Lambda\gamma_e} \qquad 2.92b$$

(in a linear polyatomic we must consider the z-component of the vibrational or vibronic, rather than the electronic, angular momentum). The transformations

$$\hat{C}_z(\theta) : \gamma_e \to \gamma_e + \theta \qquad 2.93a$$

and

$$\hat{\sigma}(yz) : \gamma_e \to -\gamma_e \qquad 2.93b$$

lead to the induced action

$$\hat{C}_z(\theta) : |\Lambda_e\rangle \to |\Lambda_e\rangle\, e^{-i\Lambda_e\theta} \qquad 2.94a$$

$$\hat{\sigma}(yz) : |\Lambda_e\rangle \to |-\Lambda_e\rangle \qquad 2.94b$$

and the reduction

$$|\Lambda_e = 0\rangle \equiv \Sigma^+ \text{ in } C_{\infty v} \qquad 2.95a$$

(Σ^- possible for a many-electron function)

and

$$\{|\Lambda_e\rangle\,,\,|-\Lambda_e\rangle\} \equiv E_{|\Lambda_e|} \text{ in } C_{\infty v}\,. \qquad 2.95b$$

Now, the product spatial molecular wavefunction must transform as the identity representation Σ^+ in $C_{\infty v}$, because all point group operations result in the identity permutation of nuclei. Allowed species in $(E,\mathcal{J}) \times C_{\infty v}$ are therefore $(\pm)\Sigma^+$, and this has the following consequences ([109] p.237):

(i) Electronic state Σ^+

For given j, the only allowed rotational wavefunction is the k =0 component with parity $\pi = (-)^j$ (physically, it is obvious that rotation of the nuclei alone cannot produce a component of angular momentum along the internuclear axis, hence k = 0). Here we have the simplest case, where the rotor has no 'internal structure'

[83], and the molecular parity is $(-)^j$ as for the usual spherical harmonics Y^j. The first few rotational levels are shown in Figure 5, where we give the symmetry labels of molecular states in $(E,\mathcal{J}) \times C_{\infty v}$ ([109] Figure 61a).

(ii) Electronic state Σ^-

For given j, the allowed rotational wavefunction is the k = 0 component with $\pi = (-)^{j+1}$. The rotational levels are shown in Figure 2.5b ([109] Figure 114a).

(iii) $\Lambda(\Lambda_e) \neq 0$

Since we must have (cf. 2.99a, 2.104a)

$$\exp(-i\theta(k + \Lambda_e)) = 1 \qquad \text{for all } \theta, \tag{2.96a}$$

$$k + \Lambda_e = 0 \quad , \tag{2.96b}$$

i.e., the component of the total and rotational angular momentum along the $\hat{\underset{\sim}{f}}_z$-axis is entirely electronic (vibronic in linear molecules).

Consider the eigenfunctions

$$\chi_c \sim (|j^\pi k\rangle|\Lambda_e\rangle + |j^\pi - k\rangle|-\Lambda_e\rangle) \tag{2.97a}$$

and

$$\chi_d \sim (|j^\pi k\rangle\ |\Lambda_e\rangle - |j^\pi - k\rangle\ |-\Lambda_e\rangle) \tag{2.97b}$$

of the reflection operator $\hat{\sigma}$. For the product wavefunctions to transform as Σ^+, we require that the states χ_c and χ_d have parities

$$\chi_c : \pi = (-)^j \ , \tag{2.98a}$$

$$\chi_d : \pi = (-)^{j+1} \ . \tag{2.98b}$$

As for the energies, there is a splitting (Λ-doubling) between the $\chi_{c,d}$ levels increasing quadratically with j ([109] p. 228); either the positive (χ_c) or negative (χ_d) superposition will have higher energy for all values of j. This yields the pattern of energy levels shown in Figure 2.6 for $\Lambda = 1$ (Π state), which exhibits the familiar alternation of parities ([109] Figure 61b). The states 2.98a,b are Herzberg's 'c' and 'd' states ([109] p. 239).

The lab-fixed components of the electric-dipole operator each span $(-)\Sigma^+$ in $(E,\mathcal{J}) \times C_{\infty v}$, leading to the selection rule (cf. 2.61; [109]III 181)

$$(+)\ \Sigma^+ \leftrightarrow (-)\ \Sigma^+ \tag{2.99}$$

for electric-dipole transitions. There is also an angular momentum constraint $|\Delta j| \leq 1$.

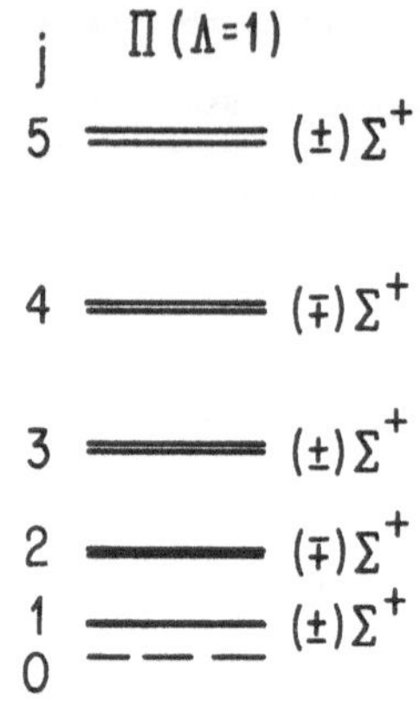

Figure 2.6 Symmetry labels for rotational levels of a heteronuclear diatomic. Π (Λ = 1) vibronic state.

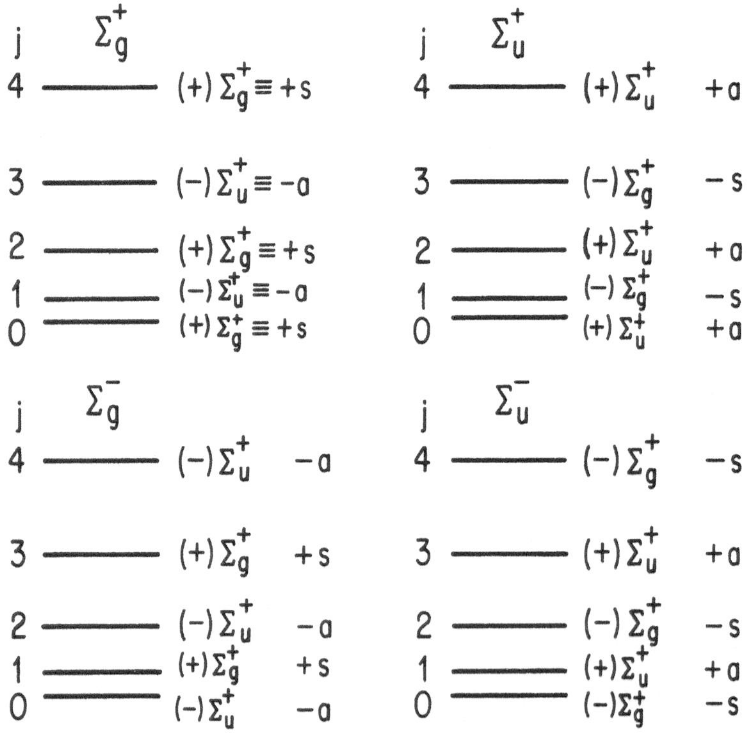

Figure 2.7 Symmetry labels for rotational levels of a homonuclear diatomic. Σ vibronic states.

(2) Homonuclear Diatomics

Nucleus 1 is now identical with nucleus 2, and the static molecular model has covering symmetry $D_{\infty h}$. Operations in the point group induce nuclear permutations as follows

$$\{\hat{E}, \hat{C}_z(\theta), \hat{\sigma}(yz)\} \rightarrow E \qquad 2.100a$$

$$\{\hat{i}, \hat{i}\,\hat{C}_z(\theta), \hat{C}_x(\pi)\} \rightarrow (12) \qquad 2.100b$$

The induced action of the inversion $\hat{i} \in D_{\infty h}$ upon the molecular coordinate is

$$\hat{i} : C \rightarrow -C \ ; \quad r_i^{\varepsilon} \rightarrow -r_i^{\varepsilon} \qquad 2.101$$

(<u>NB</u>: contrast the action of the <u>external</u> inversion

$$\mathcal{J} : C \rightarrow -C \ ; \quad r_i^{\varepsilon} \rightarrow + r_i^{\varepsilon} \qquad) .$$

We assume the electronic wavefunctions to have well-defined internal parities.

The homomorphism 2.90 from $D_{\infty h}$ onto $\{E,(12)\}$ requires that the symmetry species of the total spatial molecular wavefunction be either Σ_g^+ or Σ_u^+ in $D_{\infty h}$. From 2.90b, we note that ([109] p. 130)

$$\Sigma_g^+ \equiv \text{symmetric under exchange of nuclei} \equiv \text{'s'} \qquad 2.102a$$

$$\Sigma_u^+ \equiv \text{antisymmetric under exchange of nuclei} \equiv \text{'a'} \qquad 2.102b$$

The behaviour of the product spatial wavefunction under exchange of nuclei is therefore determined by the <u>internal</u> parity label g,u. The (external) parity of the rotational wavefunction is determined by transformation properties under $C_{\infty v}$, as before. The g,u character of the rovibronic function is then given as the product of the parities of the rotational and electronic wavefunctions. In Figure 2.7, we show rotational levels for the Σ electronic states ([109] Figure 62).

Electric-dipole selection rules are ([109] p. 130)

$$(+)\ \Sigma_g^+ \leftrightarrow (-)\ \Sigma_g^+ \qquad 2.103a$$

$$(+)\ \Sigma_u^+ \leftrightarrow (-)\ \Sigma_u^+ \ . \qquad 2.103b$$

The following correlations are required when calculating the nuclear spin statistical weights of rovibronic functions [109]:

$$\text{Bosons} \quad \left.\begin{matrix} \Sigma_g^+ \\ \Sigma_u^+ \end{matrix}\right\} \begin{matrix} \text{rovibronic function} \\ \text{associated with} \end{matrix} \left\{\begin{matrix} \text{Symmetric} \\ \text{Antisymmetric} \end{matrix}\right\} \begin{matrix} \text{nuclear} \\ \text{functions} \end{matrix}$$

$$\text{Fermions} \quad \left.\begin{matrix} \Sigma_g^+ \\ \Sigma_u^+ \end{matrix}\right\} \begin{matrix} \text{rovibronic function} \\ \text{associated with} \end{matrix} \left\{\begin{matrix} \text{Antisymmetric} \\ \text{Symmetric} \end{matrix}\right\} \begin{matrix} \text{nuclear} \\ \text{functions} \end{matrix} .$$

Chapter 3. The Symmetry Properties of Nonrigid Molecules

In this chapter we consider the theory of the symmetry properties of nonrigid molecules (NRMs). This problem has been widely discussed in recent years, with the interpretation and fundamental status of various approaches being examined very closely. Such continued interest reflects and is inspired by the lively controversies that have arisen within the subject, and the fact that some significant differences of opinion still remain.

We begin with a brief and very selective survey, which should however serve to outline those problems of particular interest to us.

As is very well known, the analysis by Longuet-Higgins [1] provides the basis for almost all subsequent work on NRMs. Following Hougen's classic work on the relation between point group operations, permutations of nuclei, and their induced action on the Born-Oppenheimer coordinates of rigid molecules (Chapter 2), Longuet-Higgins examined the fundamental symmetries of the molecular Hamiltonian (§2.1) and announced the following principle: the symmetry groups appropriate for the description of NRM energy levels and wavefunctions are (not necessarily proper) subgroups of the CNPI group (2.15). This very important idea succeeded both in rationalizing previous approaches and opening the way to a systematic study of the symmetry properties of NRMs.

As mentioned in previous chapters, the key notion here is the intuitive yet exceptionally fruitful concept of feasibility: those permutations or permutation-inversions of nuclei that can occur within the timescale of experimental observation are said to be feasible, and the PI group (cf. §2.2) or Molecular Symmetry Group (MSG) [96] is the set of all such feasible transformations. Berry [61] has emphasized that the introduction of the PI group corresponds to a process of induction from the molecular point group to an empirically determined supergroup, generated from the point group by a finite set of feasible generators [112]. Following Berry, we may refer to the supergroup in all its various guises (PI group [1], MSG [96], Q-group [68], Isodynamic group [123], Isometric group [48], group of feasible perrotations [56], symmetry group of the molecular model [49]) simply as the 'heuristic' group.

Having established the essential concepts involved in the extension of the notion of molecular symmetry to NRMs, Longuet-Higgins obtained the PI groups and associated character tables for a few typical NRMs, and proceeded as Hougen had done for rigid molecules; that is, molecular rovibronic states were classified under the PI groups, selection rules derived for, e.g., electric-dipole transitions, and nuclear spin statistical weights determined for molecular states.

These techniques have recently been described in detail by Bunker [97], and many straightforward applications and extensions of the method are collected in the review article [96]. Of the large number of specific problems that have been considered, we would mention the following important examples:

i) Hougen and Bunker's treatment of dimethylacetylene and related molecules [51-53]. This classic work on the description of nuclear motions in NRMs with low barriers to internal rotation introduced some of the novel features resulting from the parameter dependence of the FG-formalism for vibrations (cf. §1.4; also [126, 183]), and considered the transformation properties of internal molecular coordinates in the NRM symmetry group.

ii) The account by Hougen et al. [127] of the symmetry properties and dynamics of NRMs of the type $CXY_2.C \equiv C.CXY_2$. These authors gave a detailed treatment of the induced action of PI group operations upon molecular coordinates, and discussed various possible choices for the semi-rigid molecular model.

iii) Watson and Merer's account [128] of the symmetry properties of XY_2-XY_2 type molecules undergoing internal rotations. Of particular interest here is the fact that degenerate vibronic states were considered: it was noted that removal of vibronic degeneracies by internal rotation is entirely analogous to Jahn-Teller splitting in electronic states.

iv) In their work on the molecular beam electric resonance spectra of hydrogen-bonded hydrogen fluoride dimers, Dyke et al. [129] determined the PI group for the $(HF)_2$ molecule (isomorphic with the 4-group). This is an interesting example of a weakly-bound Van der Waals system treated using PI group theory. Recently, Dyke [130] has given the PI symmetry group for the hydrogen-bonded water dimer $(H_2O)_2$, which has 16 elements and is isomorphic with D_{4h} . In both cases, the rupture and re-formation of the hydrogen bond is taken to be a feasible operation. (See also the general theory for dimers in [188].)

v) Both Quack [131] and Chiu [132] have considered the correlation of rovibronic states of reactants and products in chemical reactions using PI theory, and many further applications in this area may be expected.

In view of the impressive body of work represented above, it might seem that there remained nothing more to say concerning the fundamental aspects of the subject. We consider, however, that this is not the case.

It should first be remarked that the rather intuitive nature of the original formulation [1] has perhaps been responsible for various conceptual difficulties and misunderstandings that have arisen since; as we have already seen in Chapter 2, even in the ostensibly well understood area of rigid molecule symmetry, there are some problems persisting up to the present. At the very least, then, one might hope to be able to derive a more rigorous and systematic formulation of the implementation of PI theory for particular classes of NRMs, perhaps along the lines of Louck and Galbraith's analysis of rigid molecule symmetry (cf. also Pedersen's remarks in [23]). It turns out that, despite the universal applicability of the PI approach itself, a general analysis is in fact only possible for a few rather simple and familiar types of nonrigidity, corresponding to the use of a semi-rigid molecular model (see below).

Nevertheless, it is clearly important to elucidate the generalities, such as they are, of the determination of the induced action of PI group operations upon internal NRM coordinates. The particular choice of constraints used to define the orientation of the molecule-fixed coordinate frame (e.g. Eckart-Sayvetz) is expected to be significant, as it is for rigid molecules. We stress that the PI formalism is of no direct use there, based as it is upon the fundamental symmetries of the molecular Hamiltonian, but provides a framework with which any more detailed account must necessarily be consistent.

For previous work on this general problem we refer to Hougen's discussion of normal modes in NRMs [133], Pedersen's interpretation of PI group operations [23], the work of Günthard et al. [48,134,135] based on the isometric group approach (see below), and the articles by Natanson and Adamov [126,149,154,183].

A study of the symmetry properties of individual NRMs in the PI theory soon leads to the appearance of large (typically 10-10^3 elements) and relatively unfamiliar groups [1,96,136] . This aspect of the theory then opens up the following interesting question: what, if any, are the characteristic structural features of these large NRM groups?

The search for methods to systematize and simplify the group theory has led to several investigations of the possibility of formulating NRM groups as semi-direct products [48,123-25,135,151] .

It is of interest to note that McIntosh suggested some time ago that nuclear vibrations in a sandwich structure such as dibenzene chromium could profitably be analyzed by semi-direct product methods [146,147]; our derivation of the character table for the symmetry group of the NRM dibenzene chromium (G^{+}_{144} , cf. Chapter 4) using such techniques is a confirmation of this idea [189].

However, the first serious treatment of possible semi-direct product structure in NRM symmetry groups was given by Altmann [123], who introduced the notion of isodynamic operations in this context. Altmann's work represents an attempt at a constructive approach to the heuristic supergroup starting from the molecular point symmetry group. Further discussion of the isodynamic concept is given by Altmann [124] and Watson [137], and we shall return to this topic in §3.3. A complete account of the theory of the irreducible representations of semi-direct product groups is given in [125].

The theme of semi-direct product structure has also been pursued by Woodman [138] and Serre [139], and such work suggests many possible applications to the systematic classification of NRM symmetry groups, nomenclature of normal modes, and correlation with rigid molecule symmetries etc. (cf. [126]).

The isometric group approach to NRM symmetry developed recently by Günthard et al. [48,134,135] , which is based upon the symmetry properties of the molecular graph, provides the means to tackle many of the general problems outlined above. For instance, Günthard et al . have proposed a theorem concerning semi-direct product structure of the isometric group [135]: the full isometric group is always a semi-direct product of the molecular point group and the internal isometric group [48], where the covering group is the invariant subgroup. The isometric approach is discussed in §3.4, where a counterexample to the above theorem is presented.

Finally, we would again mention Dalton's Q-group approach [68,69] (cf. §1.5). Soon after the treatment of the NRM problem by Longuet-Higgins, Dalton provided an account of PI theory [68] in which the idea of feasibility was related to the magnitude of certain matrix elements representing tunneling between localized rigid structures. The heuristic supergroup is here designated the Q-group, and is generated from the point group by permutations of nuclei corresponding to direct transitions between localized structures. The method has been extended to the case where the localized structures involved have distinct geometrical shapes [34]. Although the overall nuclear dynamics are implicit in the choice of generators for the Q-group, it is clear that the method, by adopting a matrix representation of the Hamiltonian, avoids a detailed description of the nuclear motions in the nonrigid system, and is really only appropriate where the large-amplitude motions are slower than overall rotation of the molecule [61].

With all these points in mind, we now outline the work to be described in the remainder of this chapter. We aim to present a unified approach to the symmetry properties of a particular class of NRMs, those for which the dynamical picture introduced by Sayvetz [10], and hence the use of a semi-rigid molecular model, is appropriate (§1.4). A timescale separation between slow, large-amplitude internal (or contortional [97]) motions and rapid, small-amplitude vibrations is therefore assumed in first approximation. Most detailed theoretical studies of nuclear motion in NRMs are investigations of the properties of Hamiltonians for systems of this type [5]; such work is complemented by more abstract induced representation approaches [118] like the Q-group theory.

The symmetry group of the SRMM, which we denote $\mathbf{H}$, is identified as the NRM symmetry group, and transformations of internal molecular variables induced in a well-defined fashion by the elements of $\mathbf{H}$ result in feasible permutations of nuclei. The possibility of formulating the symmetry group of the molecular model as a semi-direct product is examined, and a unified framework provided for the previous treatments by Altmann, Woodman and Günthard et al.

Isodynamic operations are defined as straightforward generalizations of the vibrational symmetry operations for rigid molecules considered in Chapter 2, and are shown to generate the invariance group of the Eckart-Sayvetz frame. The isodynamic theory therefore suppresses the equivalent rotations ($\epsilon\ O^f(3)$) by viewing induced transformations of Born-Oppenheimer coordinates from the molecule-fixed frame. Our isodynamic operations are identified with the perrotations discussed by Gilles and Philippot [56] .

The relation of our formalism to previous approaches is considered. It is not our purpose to compare various approaches directly (cf. [140]); rather, we wish to show that previous approaches can appear as particular aspects of a unified formalism, bearing in mind that a treatment of NRM symmetry in terms of any but the simplest of dynamical models [141,148] must necessarily be consistent with the PI theory.

Günthard et al. have discussed the connection between the PI and isometric theories in a similar spirit [135]. The penetrating formal analysis by Gilles and Philippot [56] should also be mentioned here. We consider, however, that several of their conclusions are rendered erroneous by a misinterpretation of the Eckart-Sayvetz conditions, and will return to this point below. The Eckart-Sayvetz conditions are of particular importance in our formalism, which may therefore be regarded as an extension of the ideas of Louck and Galbraith [17].

This chapter is based upon our previously published accounts [49, 184]. We should like to note that many of the points dealt with below were made earlier and independently in the work of Natanson and Adamov, which appeared in the Russian literature [39, 142, 149, 183] (cf. also [126, 154]). We are greatly indebted to Dr. Natanson for many stimulating communications and discussions concerning the theory of molecular symmetry.

3.1 The Symmetry Group of the Semi-rigid Molecular Model

The semi-rigid molecular model (SRMM) was introduced in Chapter 1 as a basis for describing the nuclear motions in NRMs, and we refer to §1.4 for detailed discussion.

Let us recall that the SRMM is a set of triples

$$\mathcal{A} \equiv \{\underset{\sim}{a}^{\alpha}(\gamma), m_{\alpha}, z_{\alpha}); \alpha = 1...N\} \tag{3.1a}$$

abbreviated to

$$\mathcal{A} \equiv \{\underset{\sim}{a}^{\alpha}(\gamma); \alpha = 1...N\} \tag{3.1b}$$

defining N vectors $\underset{\sim}{a}^{\alpha}(\gamma)$ as functions of a set of parameters collectively denoted γ, where each vector $\underset{\sim}{a}^{\alpha}(\gamma)$ is associated with a nucleus of mass m_{α} and charge z_{α}. γ is a representative point in the parameter domain Γ, and we shall suppose that there are $T(\lesssim 3N-6)$ independent large-amplitude curvilinear coordinates, so that γ stands for the vector $(\gamma_1, ..\gamma_t...\gamma_T)$. Each γ_t has a finite range defined in some suitable fashion.

The 3N components of the SRMM vectors

$$a^{\alpha}(\gamma)_i \equiv \hat{\underset{\sim}{e}}_i \cdot \underset{\sim}{a}^{\alpha}(\gamma) \quad i = x,y,z \tag{3.2}$$

are defined by introducing a right-handed coordinate frame $\{\hat{\underset{\sim}{e}}_i\}$ into the SRMM; $\{\hat{\underset{\sim}{e}}_i\}$ may be a principal-axis frame for a particular value of γ. From a dynamical point of view, it is convenient, but not essential [127], that the centre of mass condition

$$\sum_{\alpha} m_{\alpha} a^{\alpha}(\gamma)_i = 0 \qquad \forall \gamma; \; i = x,y,z \tag{3.2'}$$

be satisfied.

In this chapter, we illustrate our discussion of the symmetry properties of SRMMs with two particular molecular models (further examples are given in Chapter 4).

First consider the XY_3-invertor model shown in Figure 3.1. The SRMM vectors are given as functions of the single inversion angle γ, where $-\pi/2 < \gamma < \pi/2$, and the model is intended to describe inversion motions in molecules such as NH_3 [54]. In particular, the inversion transformation

$$\tau : \gamma \to -\gamma \qquad 3.3$$

is assumed to be feasible, and therefore built into the SRMM. We do not necessarily impose the condition that the bond lengths $|\underset{\sim}{a}^X(\gamma) - \underset{\sim}{a}^{Yi}(\gamma)|$ be constant throughout the inversion motion; structure relaxation is therefore allowed, and the only constraints are provided by the fundamental symmetries of the (assumed) potential-energy surface. The SRMM is however assumed to have C_{3v} point symmetry for all values of γ , one σ_v plane always coinciding with the xz-plane as shown.

The second molecular model is shown in Figure 3.2, and represents internal rotation in molecules such as CH_3-NO_2[143]. There is once again a single parameter, the torsional angle γ , which has range $0 \leq \gamma \leq 2\pi$ and measures the extent of internal rotation about the C-N axis. The feasibility of internal rotations is thus built into the model at the start. We choose to fix the coordinate axes $\{\hat{\underset{\sim}{e}}_i\}$ into the SRMM as shown in Figure 3.2, so that the $-NO_2$ group is the 'frame' while $-CH_3$ is the 'top'. For simplicity, the frame may be assumed to have C_{2v} symmetry and the top to have C_{3v} symmetry throughout the internal rotation; however, it is in fact only necessary that the feasible transformations

$$\tau_1 : \gamma \to \gamma + 2\pi/3 \qquad \underline{\text{mod}}\ 2\pi \qquad 3.4a$$

$$\tau_1^2 : \gamma \to \gamma + 4\pi/3 \qquad \underline{\text{mod}}\ 2\pi \qquad 3.4b$$

result in nuclear configurations congruent with the original.

Consider now a transformation h of SRMM vectors, defined as an ordered pair (ρ,τ)

$$h \equiv (\rho,\tau): a^\alpha(\gamma)_i \to R(\rho)_{ij} a^\alpha(\tau^{-1}(\gamma))_j \qquad 3.5$$

where ρ is a rotation or rotation-inversion of the molecular model, and τ is a mapping of the parameter space onto itself

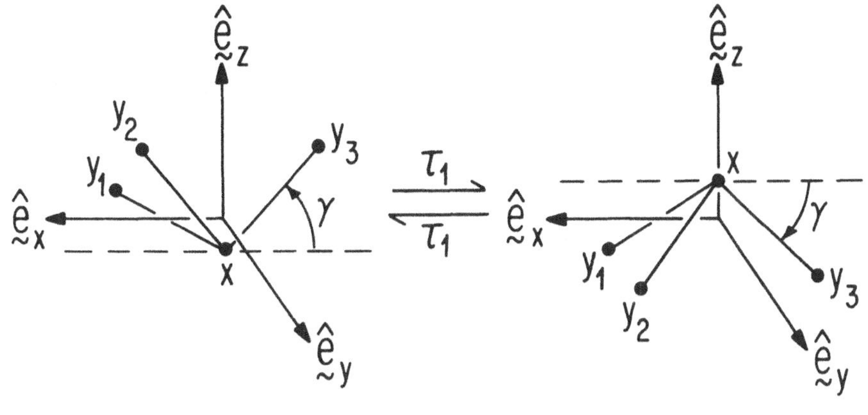

Figure 3.1 The XY_3-invertor molecule model

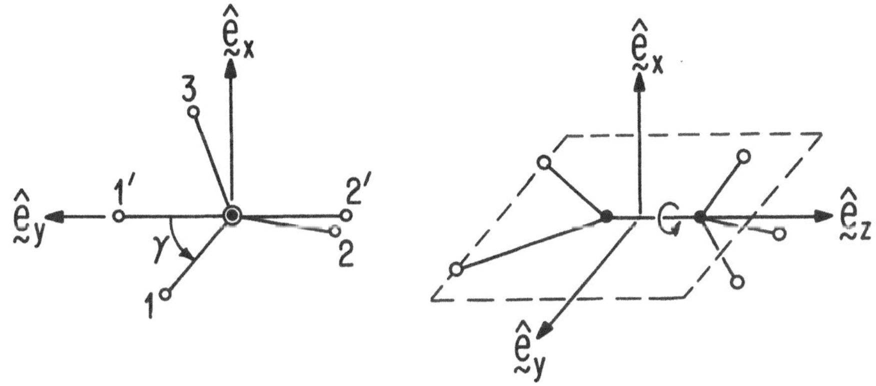

Figure 3.2 The CH_3 - NO_2 molecular model

$$\tau: \gamma \rightarrow \tau(\gamma) \varepsilon \Gamma \qquad 3.6$$

corresponding, for example, to particular changes in torsional angles etc. (cf. 3.3, 3.4). The presence of the inverse operation τ^{-1} on the right-hand side of 3.5 should be noted; this is necessary to ensure convenient results later on.

The identity transformation is denoted $h_0 = (\rho_0, \tau_0)$, where ρ_0 and τ_0 are the identity rotation and identity mapping in parameter space, respectively.

A transformation $h = (\rho,\tau)$ is termed a symmetry operation of the SRMM if it induces a permutation of SRMM vectors associated with identical nuclei; that is, if for all nuclei α

$$h=(\rho,\tau): a^{\alpha}(\gamma)_i \rightarrow R(\rho)_{ij}a^{\alpha}(\tau^{-1}(\gamma))_j = \sum_{\beta} a^{\beta}(\gamma)_i S(h)_{\beta\alpha} , \qquad 3.7$$

where S(h) is an N by N permutation matrix that permutes identical nuclei and so commutes with the diagonal mass matrix. This definition of the symmetries of the SRMM should be compared with our previous formulation of symmetry operations of the static molecular model, of which it is a straightforward generalization.

In Figure 3.3 we show the 12 symmetry operations of the XY_3-invertor model of Figure 3.1; similarly, the 12 symmetry operations of the CH_3-NO_2 internal rotation SRMM of Figure 3.2 are shown in Figure 3.4 [1] (Notation: we write $\rho_0 \equiv \hat{E}$, and $(\rho,\tau) \equiv (\rho,\gamma')$ where $\tau:\gamma \rightarrow \tau(\gamma) \equiv \gamma'$). In both cases the permutation matrices S(h) can be determined directly via the definition 3.7. It should be noted that there are two distinct symmetry operations associated with a given permutation of nuclei in the XY_3-invertor model, whereas all symmetry operations in the CH_3-NO_2 model give rise to distinct permutations. In the former case, we have an example of a primitive period transformation (see below).

The set of all symmetry operations of the SRMM forms a group, denoted **H**, which we call the symmetry group of the SRMM:

$$\mathbf{H} \equiv \{h = (\rho,\tau) | h: a^{\alpha}(\gamma)_i \rightarrow \sum_{\beta} a^{\beta}(\gamma)_i S(h)_{\beta\alpha}\} \quad . \qquad 3.8$$

We point out that $(\rho,\tau) \varepsilon \mathbf{H}$ does not necessarily imply that either (ρ,τ_0) or (ρ_0,τ) an element of **H** .

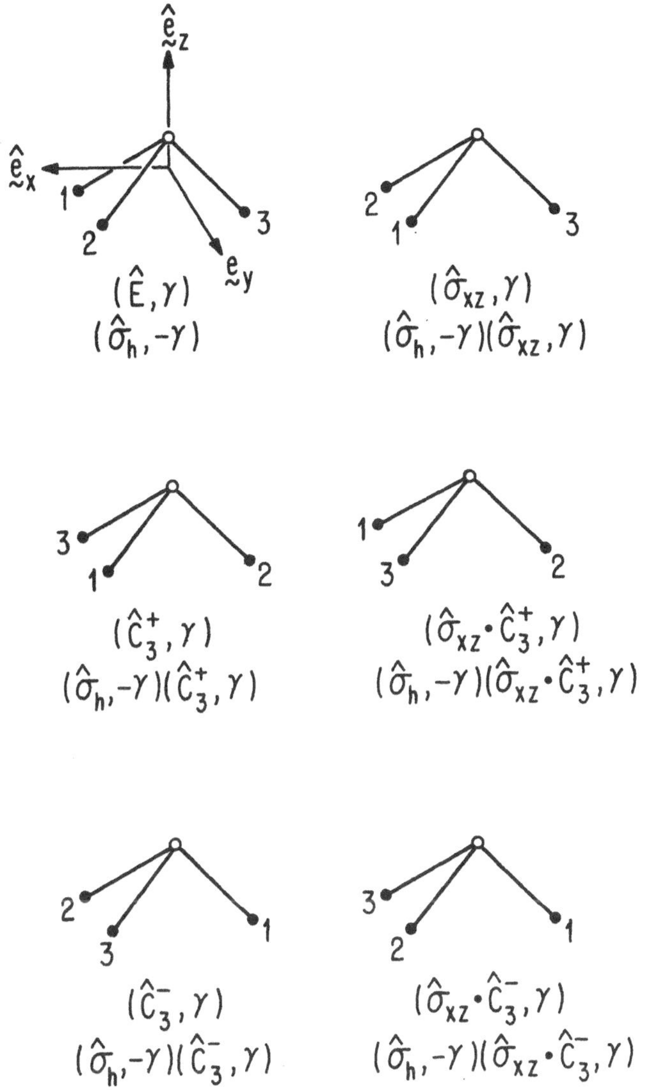

Figure 3.3 Symmetry operations for the XY_3-invertor model.

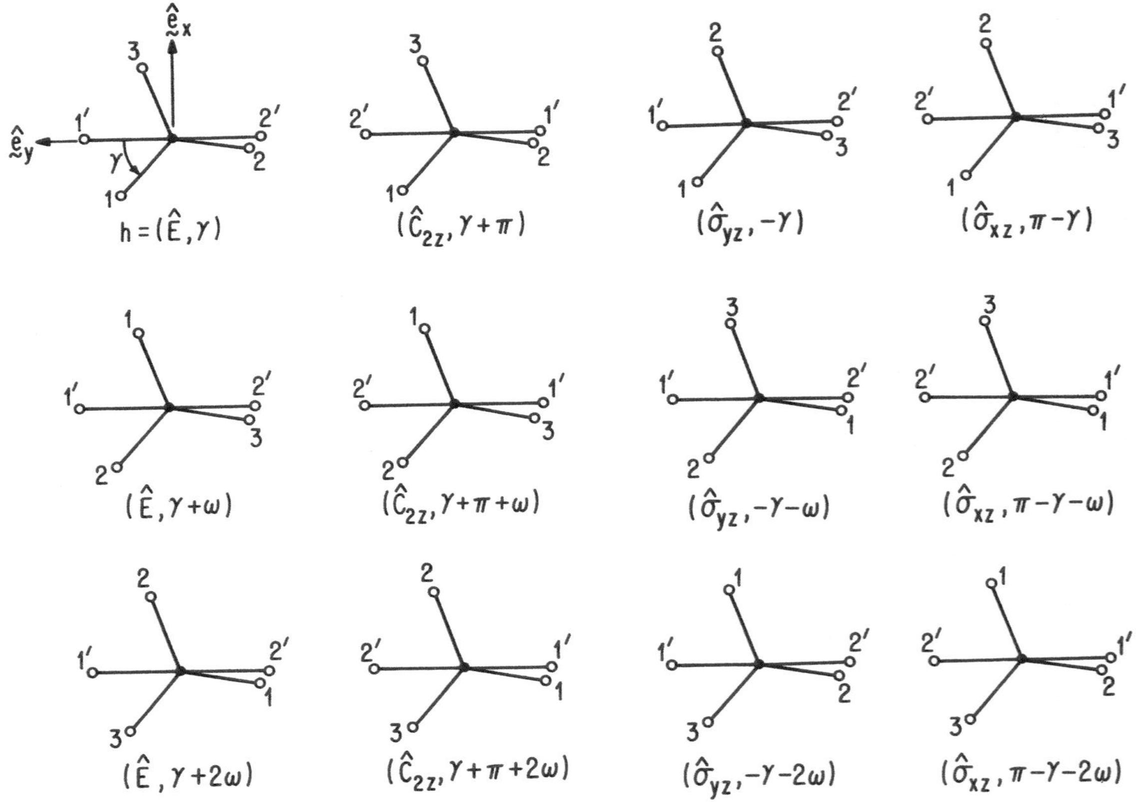

Figure 3.4 Symmetry operations for the CH_3 -NO_2 model.

Although it is of course possible to enumerate the elements of the symmetry group of the SRMM directly as we have done in Figures 3.3 and 3.4, for general purposes it is essential to establish a multiplication rule in **H** . By definition, the permutation matrices $\{S(h)\}$ then form a representation of **H**; i.e., for all elements h_1, h_2 of **H**,

$$S(h_1)S(h_2) = S(h_1 h_2) \quad . \qquad 3.9$$

The simplest and most useful multiplication rule is the direct product law

$$h_2 h_1 = (\rho_2, \tau_2)(\rho_1, \tau_1) \equiv (\rho_2 \rho_1, \tau_2 \tau_1) \in \mathbf{H} \quad . \qquad 3.10$$

It is sufficient for the validity of the simple rule 3.10 that the parametrization of the SRMM be chosen such that all the rotation and rotation-inversion operations ρ are <u>fixed</u> with respect to the coordinate frame $\{\hat{e}_i\}$, <u>independent of the particular initial point</u> γ <u>in parameter space.</u> The meaning of this condition can be clarified by reference to our two examples, for which it is valid. Thus, regardless of the particular value taken by the inversion coordinate or the torsional angle, the specifications of the associated rotations of the molecular models ($\hat{\sigma}_{xy}$, $\hat{C}_{2z}$, $\hat{\sigma}_{xz}$ etc.) remain fixed with respect to the frame vectors $\{\hat{e}_i\}$.

This requirement was discussed by Gilles and Philippot [56], who pointed out that for some molecular models (including that for CH_3-NO_2: [56] p.251) it is incompatible with the conditions

$$\varepsilon_{ijk} \sum_\alpha m_\alpha \left(\frac{\partial a^\alpha(\gamma)_j}{\partial \gamma_t}\right) a^\alpha(\gamma)_k = 0 \qquad t = 1,..T \qquad 3.11$$

which are equivalent to the vanishing of internal angular momentum in the SRMM. In order to satisfy 3.11 in the CH_3-NO_2 model, for example, it would be necessary to adopt the (internal-) axis system shown in Figure 3.5, where the NO_2- and CH_3-groups counterrotate through angles determined by the ratio of the axial moments of inertia; the specification of the reflection planes associated with symmetry operations of the SRMM would then depend upon the initial value of γ .

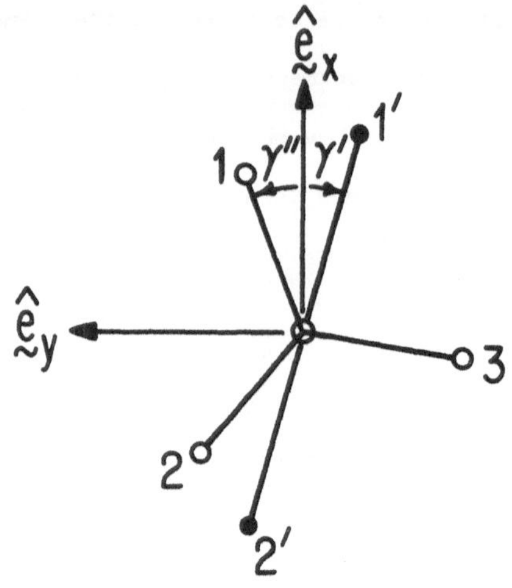

Figure 3.5 Internal-axis SRMM for CH_3 -NO_2.

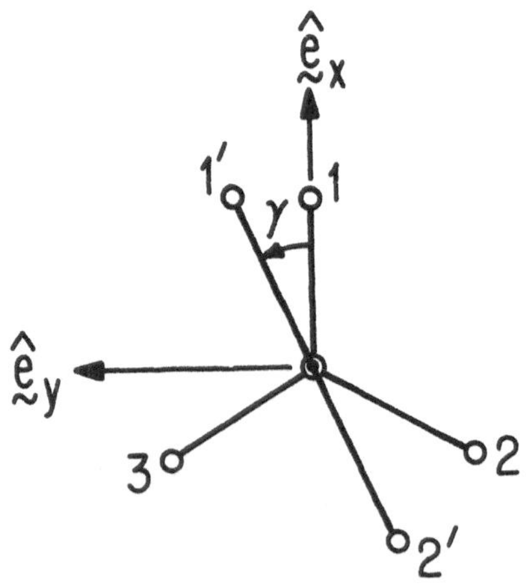

Figure 3.6 Alternative definition of the SRMM for CH_3 - NO_2, with 'top' and 'frame' interchanged.

The equations 3.11 were implicitly identified by Gilles and Philippot as the Eckart-Sayvetz conditions, and were therefore considered to be unavoidable in the theory of NRM symmetry. On this basis, it was concluded that difficulties arise concerning the formulation of isodynamic operations, which prevent the useful application of the concept ([56] pp. 240,241). We consider, however, that these arguments are incorrect. Thus, let us recall the discussion in Chapter 1 on possible definitions of the SRMM vectors $\underset{\sim}{a}^{\alpha}(\gamma)$ as functions of the parameters γ. We see that the conditions 3.11 are simply associated with the choice of an 'internal-axis' type system [5,50], and that, although these constraints uncouple internal motions and overall rotations of the molecule in zeroth approximation, <u>this is a matter of convenience rather than necessity</u>, and we are in fact free to attempt to parametrize the SRMM so that the multiplication rule 3.10 will be satisfied. This turns out to be possible for the XY_3-invertor and CH_3-NO_2 models considered here, as well as for all the molecular models treated in the next chapter, and is a question of finding a suitable geometrical specification of the $\underset{\sim}{a}^{\alpha}(\gamma)$.

Using the rule 3.10, we find that **H** is isomorphic with D_{3h} for the XY_3-invertor

$$XY_3 \qquad \mathbf{H} \overset{\text{iso}}{=} D_{3h} , \qquad 3.12$$

and for the CH_3-NO_2 model **H** (otherwise known as G_{12}) is also isomorphic with D_{3h}

$$CH_3\text{-}NO_2 \qquad \mathbf{H} \equiv G_{12} \overset{\text{iso}}{=} D_{3h} . \qquad 3.13$$

These (quite small) groups will be used to illustrate some general considerations on the structure of NRM groups; it must however be borne in mind that NRM symmetry groups are not in general isomorphic with point groups (cf. Chapter 4).

In general, two or more elements of **H** may correspond to the same permutation of SRMM vectors. If this is the case, then there exist one or more nontrivial elements of **H** corresponding to the identity permutation

$$h = (\rho,\tau) \text{ with } \rho \neq \rho_0 \text{ such that } h \neq h_0 \text{ and } S(h) = 1 \qquad 3.14$$

It should be noted that we only consider elements with $\rho \neq \rho_0$. Those with

$\rho = \rho_0$ are transformations (ρ_0,τ) corresponding to the identity permutation and acting only on the parameter space Γ ; in other words, they define the boundaries of Γ (for example, increase a torsional angle by 2π ; cf. equation 3.4) and so do not appear in **H** (all the h must be taken modulo these elements).

We refer to the operations defined by 3.14 as primitive period transformations (PPTs), and show below that they correspond to the primitive period isometric transformations introduced by Günthard et al. [135]. In the XY_3-invertor model, the 2 elements

$$(\hat{E},\gamma) \ , \ (\hat{\sigma}_h, -\gamma)$$

both correspond to the identity permutation, so that the transformation $(\hat{\sigma}_h, -\gamma)$ is a PPT. There are no PPTs for CH_3-NO_2.

Characterization of the PPTs via equation 3.14 enables us to obtain some useful general results rather easily. Thus, the set of PPTs together with the identity h_0 forms a subgroup of **H** , as can be verified from the definition 3.14 and equations 3.9 and 3.10. Moreover, using the representation property of the permutation matrices, it follows that the subgroup of PPTs is in fact an invariant subgroup of **H**, and there is a homomorphism from **H** onto the group (denoted $\mathbf{H}^{\pi}$) of distinct feasible permutations

$$\mathbf{H} \xrightarrow{\text{ho}} \mathbf{H}^{\pi} \qquad 3.15$$

where the PPTs form the kernel of the homomorphism. In the XY_3 model, it is easily verified that the group generated by the PPT is an invariant subgroup of D_{3h}

$$\{(\hat{E},\gamma) \ , \ (\hat{\sigma}_h, -\gamma)\} \qquad \mathbf{H} \overset{\text{iso}}{=} D_{3h} \quad . \qquad 3.16$$

It must be stressed that the existence of PPTs depends upon our choice of SRMM parameters. Thus, for the reasons mentioned above in connection with the multiplication rule 3.10, it is necessary for a molecule such as dimethylacetylene (which has two identical ends) to adopt a parametrization that does not distinguish a 'top' end from a 'frame' end. It is well known that the resulting PPT is associated with the use of a 'double' PI group in the description of molecular vibrations [51,52].

Although we consider the isometric group in more detail in §3.4, we note here that our development of the theory displays clearly the basis for the isometric approach to NRM symmetry [48]. Thus, it can be seen from equations 3.5 and 3.7 that a transformation of internal parameters of the form

$$\hat{P}_\tau : a^\alpha(\gamma)_i \to a^\alpha(\tau^{-1}(\gamma))_i \quad , \qquad 3.17$$

where $(\rho,\tau) \in \mathbf{H}$ for some ρ , merely permutes the elements of the distance set Δ , where

$$\Delta \equiv \{|\underset{\sim}{a}^\alpha(\gamma) - \underset{\sim}{a}^{\alpha'}(\gamma)|;\ \alpha > \alpha' = 1,..N\} \qquad 3.18$$

(so as to induce an automorphism of the vertex-valued molecular graph), and is called an <u>internal isometric transformation.</u> The group of all distinct internal isometric transformations is denoted $\overline{\overline{\mathcal{F}}}$ (following [135])

$$\overline{\overline{\mathcal{F}}} \equiv \{\tau|(\rho,\tau) \in \mathbf{H}\} \qquad 3.19$$

and there is in general a homomorphism from $\mathbf{H}$ onto the internal isometric group

$$\mathbf{H} \xrightarrow{\text{ho}} \overline{\overline{\mathcal{F}}} \qquad 3.20$$

The problem of the structure of the isometric group is taken up in §3.4.

We now show that the PPTs defined above correspond to the primitive period isometric transformations introduced in [135]. For, if $h = (\rho,\tau)$ is a PPT according to 3.24, then

$$a^\alpha(\tau^{-1}(\gamma))_i = \tilde{R}(\rho)_{ij} a^\alpha(\gamma)_j \qquad 3.21$$

so that for all α, α'

$$|\underset{\sim}{a}^\alpha(\tau^{-1}(\gamma)) - \underset{\sim}{a}^\alpha(\tau^{-1}(\gamma))| = |\underset{\sim}{a}(\gamma) - \underset{\sim}{a}^\alpha(\gamma)| \qquad 3.22a$$

with

$$\underset{\sim}{a}^\alpha(\tau^{-1}(\gamma)) \neq \underset{\sim}{a}^\alpha(\gamma) \qquad 3.22b$$

for some α . These are precisely the conditions defining the primitive period internal isometric transformations.

Let us consider the transformations of internal molecular variables induced by elements of the symmetry group of the SRMM. First, we recall the relation between the Born-Oppenheimer variables and the lab-fixed cartesian coordinates of nucleus α

$$R_i^\alpha - R_i = C_{ij}(a^\alpha(\gamma)_j + d_j^\alpha) \qquad 3.23$$

and electron ε

$$R_i^\varepsilon - R_i = C_{ij}r_j^\varepsilon \qquad 3.24$$

together with the T+6 (linear) constraints on the 3N components of the nuclear displacements $\{d_i^\alpha\}$

Center of mass $\quad \sum_\alpha m_\alpha d_i^\alpha = 0 \qquad i = x,y,z, \qquad 3.25a$

Eckart $\quad \varepsilon_{ijk} \sum_\alpha m_\alpha a^\alpha(\gamma)_j d_k^\alpha = 0 \qquad i = x,y,z, \qquad 3.25b$

Sayvetz $\quad \sum_\alpha m_\alpha\left(\frac{\partial a^\alpha(\gamma)_i}{\partial \gamma_t}\right) d_i^\alpha = 0 \qquad t = 1..T. \qquad 3.25c$

Then, each transformation $h = (\rho,\tau) \varepsilon \mathbf{H}$ induces the following transformation of molecular variables (cf. 2.31)

$$h: C \to C\tilde{R}(\rho) \qquad 3.26a$$

or

$$h: C \to (\det R(\rho))C\tilde{R}(\rho) \quad , \qquad 3.26a'$$

$$h: a^\alpha(\gamma)_i \to L_h \cdot a^\alpha(\gamma)_i \equiv \sum_\beta R(\rho)_{ij} S(h)_{\alpha\beta} a^\beta(\gamma)_j \quad , \qquad 3.26b$$

$$h: d_i^\alpha \to L_h \cdot d_i^\alpha \equiv \sum_\beta R(\rho)_{ij} S(h)_{\alpha\beta} d_j^\beta \quad , \qquad 3.26c$$

$$h: R_i \to R_i \quad , \qquad 3.26d$$

$$h: r_i^\varepsilon \to C_{ij} r_j^\varepsilon \quad . \qquad 3.26e$$

These transformations generalize those given in Chapter 2 for rigid molecules, and are a realization of the group H, which is identified as the NRM symmetry group.

Elements of the NRM symmetry group therefore induce the following (not necessarily distinct) transformations of position vectors: for all elements of **H**,

$$h: \underset{\sim}{r}^{\alpha} \rightarrow \underset{\sim}{r}^{\bar{\alpha}} \equiv \sum_{\beta} S(h)_{\alpha\beta} \underset{\sim}{r}^{\beta} \qquad 3.27$$

<u>or</u>

$$h: \underset{\sim}{r}^{\alpha} \rightarrow (\det R(\rho)) \underset{\sim}{r}^{\bar{\alpha}} \quad , \qquad 3.27'$$

and

$$h: \underset{\sim}{r}^{\varepsilon} \rightarrow \underset{\sim}{r}^{\varepsilon} \qquad 3.28$$

<u>or</u>

$$h: \underset{\sim}{r}^{\varepsilon} \rightarrow (\det R(\rho)) \underset{\sim}{r}^{\varepsilon} \quad , \qquad 3.28'$$

i.e., for the unprimed choice of transformation of rotational variables 3.26a, elements of the NRM symmetry group induce permutations $(\mathcal{P}_h \in \mathbf{H}^{\pi})$ of the nuclear position vectors $\{\underset{\sim}{r}^{\alpha}\}$ (3.27,3.28), where the correspondence between symmetry operations of the SRMM and permutations is defined <u>via</u> the relation 3.7. The permutations are interpreted in exactly the same fashion as the permutations associated with point group operations in rigid molecules, considered in §2.3.

Use of the feasible transformations of rotational variables 3.26a' results in the permutations or permutation-inversions of nuclei 3.27' and 3.28', which are elements of the PI group.

The introduction of permutation-inversions enables us to distinguish between the identity h_0 and a given PPT when the rotation ρ associated with the PPT is <u>improper</u>. The PI group is then an extension of $\mathbf{H}^{\pi}$ by the inversion operation $\mathcal{J}\,(E^*)$, where $\mathcal{J}$ affects only the rotational coordinates (cf. 2.43; also [99,100]). This is analogous to the fact that the PI group for a planar rigid molecule is always isomorphic with the point group, whereas there is a 2:1 homomorphism from the point group onto the group of distinct permutations.

There are two important points to be made concerning the transformations 3.26. First, the operations ρ define the <u>equivalent rotations</u> introduced by Longuet-Higgins [1]. The transformations 3.26a (3.26a') are defined with respect to the Eckart frame and are elements of the group $O^f(3)$ ($SO^f(3)$). As already noted in Chapter 2, this conclusion seems to be contrary to the interpretation of PI theory given by Louck and Galbraith [17]. Secondly, the transformations 3.26 leave invariant the dynamical constraints 3.25 (as can be seen most easily from the least-squares criterion [39]; cf. §3.3), so that we can interpret 3.25b as a change in internal parameter

$$h: a^{\alpha}(\gamma)_i \rightarrow a^{\alpha}(\gamma')_i \qquad 3.29a$$

with

$$\gamma' = \tau(\gamma) \quad . \qquad 3.29b$$

We can therefore write the induced action of h as the mapping (cf. 2.36; [17] 6.16; [135] 5.4):

$$h: (C,\gamma,d_i^{\alpha},r_i^{\varepsilon}) \rightarrow (C\tilde{R}(\rho),\tau(\gamma),L_h\cdot d_i^{\alpha},R(\rho)_{ij}r_j^{\varepsilon}) \qquad 3.30$$

or

$$h: (C,\gamma,d_i^{\alpha},r_i^{\varepsilon}) \rightarrow ((\det R(\rho))C\tilde{R}(\rho),\tau(\gamma),L_h\cdot d_i^{\alpha},R(\rho)_{ij}r_j^{\varepsilon}) \quad . \qquad 3.30'$$

In this form, the coordinate transformations can be used to classify tensor operators and wavefunctions pertaining to rotational, internal motion, vibrational and electronic degrees of freedom in the NRM symmetry group (cf. §2.2). In particular, 3.26c provides a faithful 3N-dimensional vibrational representation

$$\{R(\rho) \otimes S(h) \,|\, h \in \mathbf{H}\} \qquad 3.31$$

which is associated with the use of 'external' cartesian displacement coordinates, and which will not necessarily be identical with the vibrational representations obtained using various sets of 'internal' symmetry coordinates [96,144].

When all interactions between molecular degrees of freedom are taken into account, the 'rotorvibronic' product spatial wavefunction must be classified in $\mathbf{H}^{\pi}$. Allowed irreducible representations of $\mathbf{H}$ are those that are totally symmetric under the PPTs, since these are all equivalent to the identity permutation of nuclei. The Pauli principle is then implemented as a restriction on the IRs of $\mathbf{H}^{\pi}$ spanned by the product of the 'rotorvibronic' and nuclear-spin functions.

It is apparent that the group $\mathbf{H}$ is a complete characterization of the NRM symmetry, and must in general be determined by a process of exhaustion based upon a study of the tranformations 3.7. Furthermore, a treatment of NRM symmetry based upon the symmetry group of the SRMM is a direct and natural generalization of the approach to rigid molecule symmetry described in §2.2.

It must be emphasized that, when formulating the elements of $\mathbf{H}$ as ordered pairs (ρ,τ), we are not just obtaining yet another realization of the associated PI or isometric group. For, given a multiplication rule such as 3.10, the elements of $\mathbf{H}$ can be manipulated directly, without any reference to the underlying permutations. This turns out to be very convenient in practice, since rotation operations (ρ) and transformations (τ) of a relatively small number of internal variables can be manipulated much more easily than can permutations of large numbers of nuclei. Also, when the symmetry group of the SRMM possesses nontrivial PPTs, the structure of $\mathbf{H}$ corresponds directly to the extended ('double','quadruple', etc.) PI group. The group $\mathbf{H}^{\pi}$ of distinct permutations is then isomorphic with the factor group of the invariant subgroup of PPTs, obtained by mapping all PPTs onto the identity. This is the reverse of the usual procedure [52], in which a suitable extension of the PI group has to be constructed.

In the next section, we investigate the possibility of constructing the symmetry group of the SRMM as a semi-direct product. We find that our formulation of $\mathbf{H}$ is particularly well-suited to describe the structural features of the NRM symmetry group that are of interest in this problem, and to take practical advantage of any such features.

3.2 The Structure of $\mathbf{H}$

Even for only moderately complicated SRMMs, the associated symmetry group can be very large. It is therefore important to be able to make full use of any internal structure that $\mathbf{H}$ may possess, when obtaining its irreducible representations and character table. As mentioned in the introduction, such a consideration has motivated several formulations of NRM symmetry groups as semi-direct products, and we now examine the problem.

Given the validity of the multiplication rule 3.10, the symmetry group of the SRMM has two invariant subgroups, the intrinsic group and the point group.

The intrinsic group G^I is the subgroup of **H** consisting of all intrinsic operations of the SRMM: these are defined as transformations acting only upon the internal parameter space, so that the associated rotation is the identity

$$G^I \equiv \{(\rho,\tau) \in \mathbf{H} | \rho = \rho_0\} \quad . \qquad 3.32$$

An example of a nontrivial intrinsic group occurs in the CH_3-NO_2 molecular model. We have (in the notation of Figure 3.4)

$$G^I = \{(\hat{E},\gamma),(\hat{E},\gamma + 2\pi/3),(\hat{E},\gamma + 4\pi/3)\} \overset{iso}{=} C_3 \qquad 3.33$$

and

$$C_3 \triangleleft \mathbf{H} \equiv G_{12} \quad , \qquad 3.34$$

as can be verified explicitly. In this case, the intrinsic operations correspond to rotations of the methyl group about the C-N bond, and clearly do not involve an overall rotation of the SRMM.

The point group G^P is defined by the covering symmetry of the SRMM for arbitrary γ . The covering operations do not act upon the parameter space

$$G^P \equiv \{(\rho,\tau) \in \mathbf{H} | \tau = \tau_0, \forall \gamma\} \qquad 3.35$$

The XY_3-invertor has a nontrivial point group G^P, which is the C_{3v} subgroup of **H** (cf. Figure 3.3)

$$G^P = \{(\hat{E},\gamma),(\hat{C}_3,\gamma),(\hat{C}_3^2,\gamma),(\hat{\sigma}_v,\gamma),(\hat{\sigma}_v\hat{C}_3,\gamma),(\hat{\sigma}_v\hat{C}_3^2,\gamma)\} \overset{iso}{=} C_{3v} \qquad 3.36$$

and

$$C_{3v} \triangleleft \mathbf{H} \quad . \qquad 3.37$$

We would stress that the point group G^P is a global property of the SRMM, in the sense that we must find the highest common symmetry over the whole of the parameter range, for only then is G^P an invariant subgroup of **H**. Although problems do not normally arise here, since molecular models are usually defined to have a given fixed (with respect to $\{\hat{e}_i\}$) covering symmetry for all values of γ except for a few isolated points, we note the subtle problems discussed by Nourse and Mislow in connection with correlated ring rotations in the tetraphenylmethane molecule

[145]. Even though the molecular model is assumed to have non-trivial S_4 covering symmetry at all times, it turns out that the group S_4 is not an invariant subgroup of the symmetry group of the molecular model (G_{384}, cf. Chapter 4). This is because there are <u>three</u> distinct S_4 groups involved, having the unique $\hat{S}_4$-axes aligned along the x-,y- and z-axes respectively, and the intersection of these groups is C_1, the trivial point symmetry. The 'point group' $\mathbf{G}^P$ associated with this model is therefore C_1.

Using the multiplication rule 3.10 it is simple to verify that both $\mathbf{G}^I$ and $\mathbf{G}^P$ are invariant subgroups of $\mathbf{H}$. Furthermore, the elements of $\mathbf{G}^I$ and $\mathbf{G}^P$ commute, and we can therefore form the direct product $\mathbf{G}^I \otimes \mathbf{G}^P$, which is also an invariant subgroup of $\mathbf{H}$

$$\mathbf{H} \rhd \mathbf{G}^I \otimes \mathbf{G}^P \ . \tag{3.38}$$

The factor group $\mathcal{K}$ may be defined

$$\mathcal{K} \equiv \mathbf{H}/(\mathbf{G}^I \otimes \mathbf{G}^P) \quad . \tag{3.39}$$

For the XY_3-invertor model, we have

$$\mathbf{G}^I \otimes \mathbf{G}^P = C_1 \otimes C_{3v} \overset{\text{iso}}{=} C_{3v} \tag{3.40a}$$

so that

$$\mathcal{K} = D_{3h}/C_{3v} \overset{\text{iso}}{=} \mathcal{V}_2 \quad , \tag{3.40b}$$

($\mathcal{V}_2$ ≡ abstract group of order 2), and for the CH_3-NO_2 model

$$\mathbf{G}^I \otimes \mathbf{G}^P = C_3 \otimes C_1 \overset{\text{iso}}{=} C_3 \tag{3.41a}$$

with

$$\mathcal{K} = G_{12}/C_3 \overset{\text{iso}}{=} C_{2v} \quad . \tag{3.41b}$$

Now, in general it is <u>not</u> possible to find a set of coset representatives of $\mathbf{G}^I \otimes \mathbf{G}^P$ that closes to form a group (denoted $\mathbf{K}$) isomorphic with the factor group $\mathcal{K}$. This assertion can be verified by considering the SRMM for molecules of the type XY_2-XY_2 undergoing internal rotation, which has a symmetry group $\mathbf{H}$($\equiv G_{16}^+$, [128]) of order 32 (cf. Chapter 4). Here,

$$G^I \text{ igo } \mathcal{V}_2 \qquad 3.42a$$

$$G^P \text{ igo } D_2 \qquad 3.42b$$

and

$$\mathcal{K} \text{ igo } C_{2v} \quad , \qquad 3.42c$$

but it is not possible to form a group of 4 coset generators of $\mathcal{V}_2 \otimes D_2$ isomorphic with the factor group C_{2v}.

If, however, it is possible to find such a group of coset generators, then H can be written as a semi-direct product ([125]; cf. also Appendix 3)

$$\mathbf{H} = (G^I \otimes G^P) \circledS \mathbf{K} \quad , \qquad 3.43$$

which can be rearranged to

$$\mathbf{H} = G^I \circledS (G^P \circledS \mathbf{K}) \qquad 3.44a$$

or

$$\mathbf{H} = G^P \circledS (G^I \circledS \mathbf{K}) \quad . \qquad 3.44b$$

Since all elements of K except the identity involve nontrivial rotations or rotation-inversions ρ, K is isomorphic with a point symmetry group. The point group $G^P \circledS \mathbf{K}$ can be thought of as the effective covering symmetry of the NRM induced by the nonrigidity. For example, the tetraphenylmethane molecule mentioned previously has effective tetrahedral (T_d) symmetry. The particular group structure 3.44a corresponds to the most recent version of Altmann's theory [125], and to the reduction given by Woodman for the symmetry groups of NRMs with internal rotation [138]. The group structure 3.44b corresponds to the abstract semi-direct product for the complete isometric group proposed by Günthard et al. [135], with

(cf. §3.4)

$$\overline{\overline{\mathcal{F}}} \text{ igo } G^I \circledS \mathbf{K} \; . \qquad 3.45$$

The symmetry groups of both the XY_3-invertor and CH_3-NO_2 SRMMs can be written as semi-direct products. For the XY_3-invertor, we have

$$\mathbf{K} \doteq \{(\hat{E},\gamma),(\hat{\sigma}_h,-\gamma)\} \text{ igo } C_h \text{ igo } \mathcal{V}_2 \qquad 3.46a$$

and

$$\mathbf{H} = C^P_{3v} \circledS C_h \quad . \qquad 3.46b$$

However, in this case **K** is also the group generated by the PPT $(\hat{\sigma}_h, -\gamma)$, and is itself an invariant subgroup of **H**. The symmetry group for the XY_3-invertor is therefore a direct product

$$H = C_{3v}^{P} \otimes C_h = D_{3h} \ . \qquad 3.47$$

Note that the symmetry group of the XY_3-invertor has the semi-direct product structure 3.46 entirely in accord with our general analysis; this result should be contrasted with Woodman's remarks on the ammonia symmetry group [138].

For the CH_3-NO_2 model, we have

$$K = \{(\hat{E},\gamma),(\hat{C}_{2z},\gamma + \pi),(\hat{\sigma}(xz),\gamma-\pi),(\hat{\sigma}(yz),-\gamma)\} \ \underline{\underline{\text{iso}}} \ C_{2v} \qquad 3.48a$$

$$H = C_3^I \circledS C_{2v} \equiv G_{12} \qquad . \qquad 3.48b$$

We would emphasize that the reduction of **H** as a semi-direct product is not necessarily unique, depending as it does upon the particular parametrization of the SRMM. Thus, consider the CH_3-NO_2 molecular model shown in Figure 3.6, in which the roles of 'frame' and 'top' have been reversed with respect to Figure 3.2. In this case

$$G^P \ \underline{\underline{\text{iso}}} \ C_1 \qquad 3.49a$$

$$G^I \ \underline{\underline{\text{iso}}} \ C_2 \qquad 3.49b$$

$$K \ \underline{\underline{\text{iso}}} \ C_{3v} \qquad 3.49c$$

and

$$H' = C_2^I \circledS C_{3v} \ \underline{\underline{\text{iso}}} \ H \equiv G_{12} \qquad . \qquad 3.50$$

A change in the choice of coordinate frame therefore induces an automorphism of the symmetry group of the SRMM, as noted by Günthard et al. [48].

Let us define an extended intrinsic group $\tilde{G}^I$ as the subgroup of **H** consisting of all transformations (ρ,τ) with either $\rho = \rho_0$ or $\rho = i$, the inversion operation. This definition is only useful provided that the point group G^P does not itself contain the inversion i:

$$\tilde{G}^I \equiv \{(\rho,\tau) \in H | \rho = \rho_0 \text{ or } \rho = i; \ (i,\tau_0) \notin G^P\} \ . \qquad 3.51$$

Since the inversion commutes with all $\rho \in O(3)$, $\tilde{G}^I$ is an invariant subgroup of H

$$H \rhd \tilde{G}^I \quad , \tag{3.52}$$

and we can still form a direct product $\tilde{G}^I \otimes G^P$, so that

$$H \rhd \tilde{G}^I \otimes G^P \quad . \tag{3.53}$$

The appropriate factor group is denoted

$$\tilde{\mathcal{K}} \equiv H/(\tilde{G}^I \otimes G^P) \quad . \tag{3.54}$$

The point of this is that while it might not be possible to form a group K isomorphic with $\mathcal{K}$, and thereby to write H as the semi-direct product 3.43, it may well be possible to find a group $\tilde{K}$ that is isomorphic with $\tilde{\mathcal{K}}$ (recall that $\tilde{\mathcal{K}}$ is smaller than $\mathcal{K}$). The group H can then be written as the semi-direct product

$$H = (\tilde{G}^I \otimes G^P) \circledwedge \tilde{K} \quad , \tag{3.55}$$

which can be rearranged to

$$H = (\tilde{G}^I \circledwedge G^P) \circledwedge \tilde{K} \quad , \tag{3.56a}$$

or

$$= G^P \circledwedge (\tilde{G}^I \circledwedge \tilde{K}) \quad . \tag{3.56b}$$

Comparison of 3.56b with equations 3.44 and 3.45 shows that introduction of the extended intrinsic group corresponds to an alternative breakdown of the internal isometric group $\bar{\mathcal{F}}$ as a semi-direct product.

As an example, we note that it is necessary to invoke an extended intrinsic group $\tilde{G}^I$ for the SRMM describing coaxial rotors of the type XY_2-XY_2 (cf. 3.42):

$$\tilde{G}^I \overset{\text{iso}}{=} C_{2i} \tag{3.57a}$$

$$G^P \overset{\text{iso}}{=} D_2 \tag{3.57b}$$

$$\tilde{\mathcal{K}} \overset{\text{iso}}{=} \mathcal{V}_2 \tag{3.57c}$$

so that H can be written as the semi-direct product

$$H = (C_2^I \otimes D_{2i}^P) \circledS C_{S'} \overset{iso}{=} G_{16}^+ \qquad 3.58$$

The structure of this and related groups is discussed further in Chapter 4.

Finally, it should be apparent that a knowledge of the component groups $G^I(\tilde{G}^I)$, G^P and K $(\tilde{K})$ in the case that H can be written as a semi-direct product 3.43 (3.55) enables us to use systematic methods (described, for example, by Altmann [125]; see also [138, 139, 147]) that are available for the construction of the irreducible representations and character table of the group **H**. Semi-direct product structure thereby provides the basis for a detailed and physically significant nomenclature for the IRs of NRM groups [138], and we shall turn to this subject in Chapter 4 (see Appendix 3).

3.3 Isodynamic Operations

In this section we define the group of isodynamic operations associated with the NRM symmetry group.

Consider once again the transformations of elements of molecule-fixed nuclear coordinates induced by elements of H (3.26b,c)

$$L_h \cdot a^\alpha(\gamma)_i = \sum_\beta R(\rho)_{ij} S(h)_{\alpha\beta} a^\beta(\gamma)_j \qquad 3.59a$$

$$= a^\alpha(\tau(\gamma))_i \quad , \qquad 3.59b$$

$$L_h \cdot d_i^\alpha = \sum_\beta R(\rho)_{ij} S(h)_{\alpha\beta} d_j^\beta \quad . \qquad 3.60$$

The crucial result is then: <u>the group of transformations</u> $L_H \equiv \{L_h | h \in \mathbf{H}\}$ <u>is the invariance group of the Eckart-Sayvetz frame.</u> We can therefore consider the set L_H of molecule-fixed transformations independently of the associated transformations of rotational variables, and we shall use the generic term <u>isodynamic operation</u> to describe transformations of this type.

To show that operations L_h leave the orientation of the Eckart-Sayvetz frame invariant, we simply note that the quadratic form in the nuclear displacements

([39]; cf. Chapter 1)

$$\mathcal{E} \equiv \sum_{\alpha} m_{\alpha} \underset{\sim}{d}^{\alpha} \cdot \underset{\sim}{d}^{\alpha} \qquad 3.61$$

is invariant under all L_h, so that transformations 3.59 and 3.60 can consistently be interpreted as producing a new reference configuration $\{\underset{\sim}{a}^{\alpha}(\tau(\gamma))\}$ together with a new set of nuclear displacements $\{\underset{\sim}{\bar{d}}^{\alpha}\}$ with respect to the initial Eckart-Sayvetz frame. This conclusion generalizes the arguments of Louck and Galbraith [17] concerning the invariance group of the Eckart frame for rigid molecules (§2.3).

The isodynamic group L_H is isomorphic with H, since the vibrational representation $\{R(\rho) \cdot S(h)\}$ is a faithful representation of the NRM symmetry group. Thus, the operation L_h involves both the rotation $R(\rho)$ and the permutation $S(h)$ associated with $(\rho,\tau) = h$, so that if two different elements $h,h' \in H$ induce identical isodynamic operations they can differ only by a factor $(\rho_0,\tau'') \in H$ corresponding to the identity permutation of nuclei. As noted earlier, transformations of this type merely serve to define the boundaries of the parameter space, and all symmetry elements $h \in H$ should be specified modulo such operations. The transformations h and h' must therefore be identical, contrary to the original supposition, and the group L_H is indeed isomorphic with H

$$L_H \overset{iso}{=} H \quad .$$

The isodynamic operations we have defined are symmetry operations of the vibration/internal-motion Hamiltonian, obtained by ignoring all interactions with molecular rotation, since the elements of L_H suppress the restorative rotations 3.26a,a'. This result has also been noted by Adamov and Natanson [149], who have stressed that every term in an expansion of the Hamiltonian in powers of the rectilinear vibrational coordinates is invariant under the homogeneous linear transformations induced by elements of the PI group - in other words, under the isodynamic operations. Vibronic interactions are easily included using the transformations of electronic variables 3.26c [128]. It is clear that the $\{L_h\}$ are a natural generalization to NRMs of the vibrational symmetry operations for rigid molecules considered in §2.3.

In Figures 3.7 and 3.8 we show the isodynamic operations induced by elements of the symmetry groups of the XY_3-invertor and CH_3-NO_2 molecular models, respectively. The 'isodynamic configurations' can be considered to have arisen from the (arbitrary) initial configuration via constrained nuclear motion with respect to the fixed Eckart-Sayvetz frame, in complete analogy with the interpretation of the Wigner symmetry operations given by Louck and Galbraith (§2.3). It is particularly striking that a straightforward implementation of the transformations L_h in the CH_3-NO_2 example leads naturally to the appearance of the 'switches' (U^I) originally introduced by Altmann [123] (hence justifying our use of the term 'isodynamic'). Our isodynamic operations have the form of the perrotations of molecule-fixed coordinates introduced by Gilles and Philippot [56], which we therefore take to be rigorously defined by the equations 3.59,3.60. Contrary to the assertion in [56], isodynamic operations do not act solely upon the equilibrium configuration of nuclei, but directly upon the instantaneous distorted configuration, as shown in Figures 3.7 and 3.8.

3.4 The Isometric Theory

In this section we study the isometric group approach to the symmetry properties of NRMs. Detailed accounts may be found in the original papers [48,134,135] and a recent review article [150].

Although it is apparent that there are similarities between the isometric theory and the formalism presented in §3.1, we would maintain that our formulation of the symmetry group of the molecular model provides insights into the structure of NRM symmetry groups which are, perhaps, obscured by the analytical complexity of the isometric approach [184].

Consider once again a transformation τ of internal parameters γ ([48], equation 3.1)

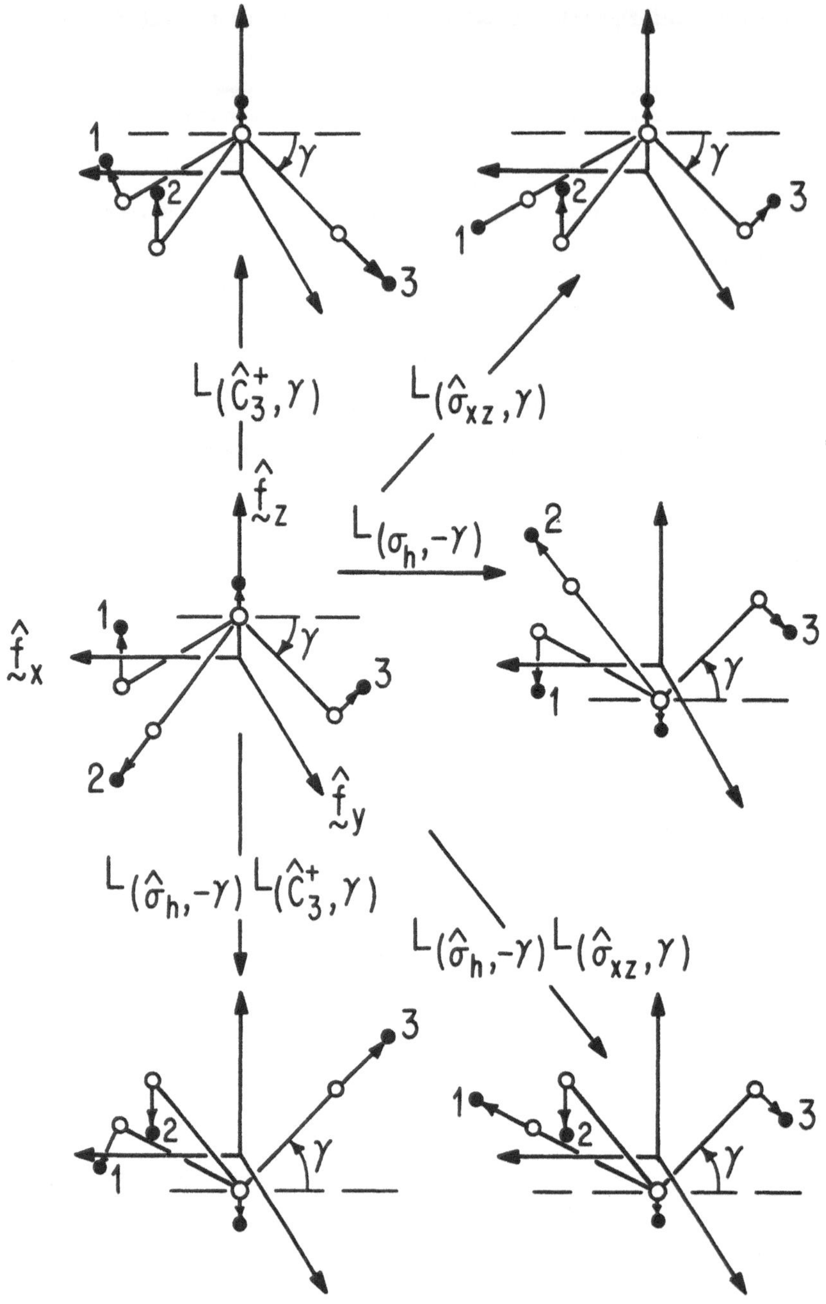

Figure 3.7 Isodynamic operations for the XY_3-invertor molecule.

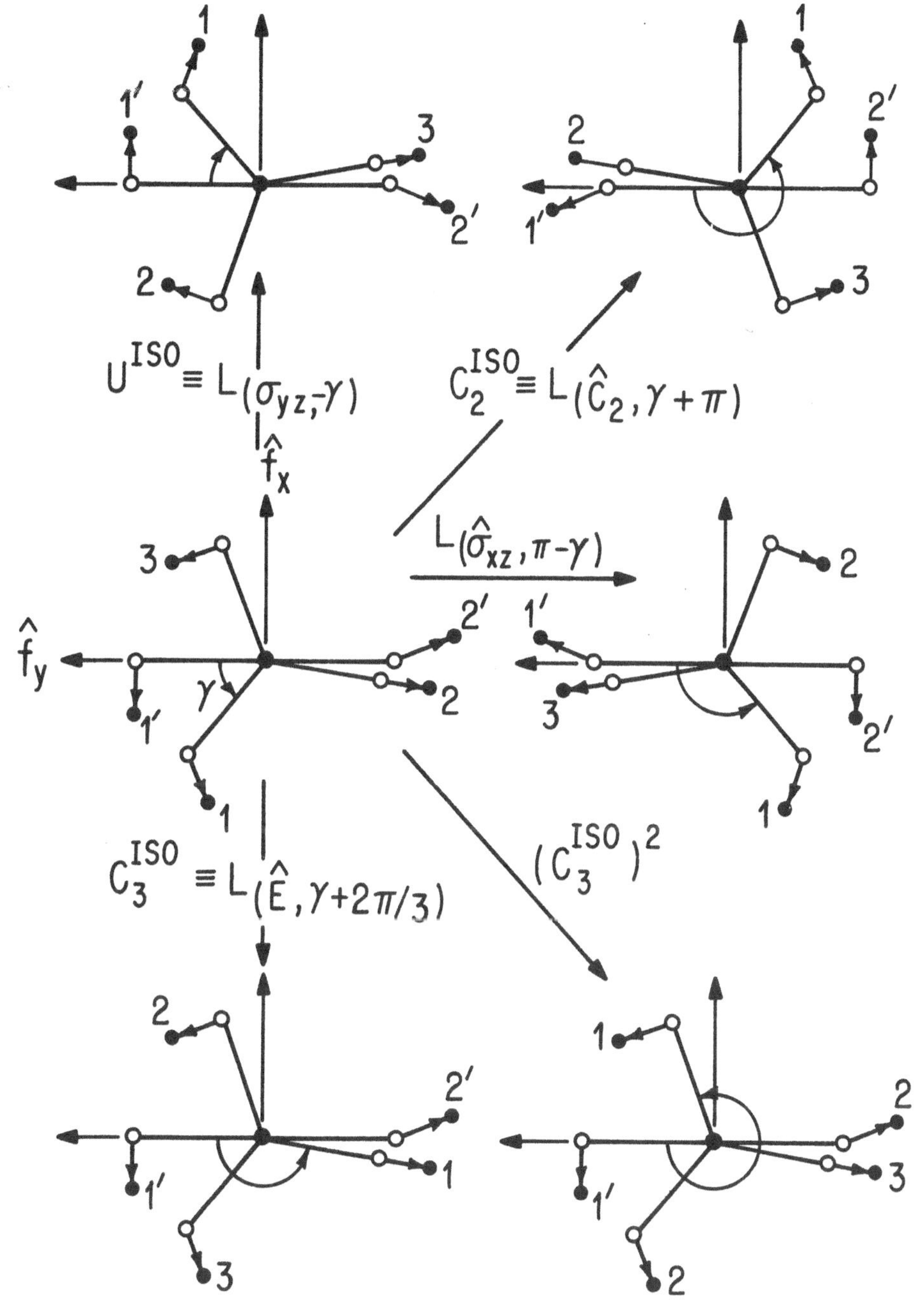

Figure 3.8 Isodynamic operations for the $CH_3 - NO_2$ molecule.

$$\tau : \gamma \rightarrow \gamma' = \tau(\gamma) \in \Gamma \quad , \tag{3.6}$$

which, although there is no necessary restriction, is usually an inhomogeneous linear transformation on the components of the vector $(\gamma_1,..\gamma_t,..\gamma_T)$ [48,135]

$$\tau : \gamma_t \rightarrow \gamma'_t = \sum_{\bar{t}} \eta(\tau)_{t\bar{t}}\gamma_{\bar{t}} + \not{t}_t \tag{3.62}$$

and has the induced action upon the SRMM vectors

$$\hat{P}_\tau : a^\alpha(\gamma)_i \rightarrow a^\alpha(\tau^{-1}(\gamma))_i \quad . \tag{3.17}$$

The transformation τ is said to be an <u>internal isometric transformation</u> if the resulting SRMM $\{\underset{\sim}{a}^\alpha(\tau^{-1}(\gamma))\}$ is related to the original $\{\underset{\sim}{a}^\alpha(\gamma)\}$ by an <u>isometry</u>: that is, if the resulting nuclear configuration can be superimposed on the original by a (finite) rotation or rotation-inversion, the net result being simply some permutation of SRMM vectors associated with identical nuclei. The set of all such transformations forms a group, the <u>internal isometric group</u> $\overline{\overline{\mathcal{F}}}$ (cf. 3.19;[48]§3.1)

$$\overline{\overline{\mathcal{F}}}(\gamma) \equiv \{\tau | \{\underset{\sim}{a}^\alpha(\gamma)\} \xleftrightarrow{\text{isometry}} \{\underset{\sim}{a}^\alpha(\tau^{-1}(\tau))\}\} \quad . \tag{3.19'}$$

According to the above definition of an isometry, for all elements of the internal isometric group it is possible to write ([48] equation 3.7; [137] equation 2.8)

$$\hat{P}_\tau : a^\alpha(\gamma)_i \rightarrow a^\alpha(\tau^{-1}(\gamma))_i = \sum_\beta a^\beta(\gamma)_j S(\tau)_{\beta\alpha} R(\tau)_{ji} \tag{3.63}$$

where $R(\tau)$ defines a finite rotation or rotation-inversion associated with the isometry, and $S(\tau)$ is an N by N matrix permuting identical nuclei. The representation of $\overline{\overline{\mathcal{F}}}$ on the nuclear configuration is denoted [135]

$$\Gamma^{(Ncf)}\{\overline{\overline{\mathcal{F}}}\} \equiv \{R(\tau) \otimes S(\tau) | \tau \in \overline{\overline{\mathcal{F}}}\} \quad . \tag{3.64}$$

Thus, in the isometric method we are explicitly interested in the set of solutions to equation 3.63, and the corresponding set of perrotations 3.64. The molecular potential energy is manifestly invariant under an isometry, and Günthard <u>et al</u>. have shown that the rotation/internal-motion kinetic energy operator is also invariant under the group of internal isometric transformations [48].

All this is quite similar to the definition we have given in §3.1 of the symmetry group of the molecular model. However, should the SRMM possess any non-trivial covering symmetry for arbitrary γ, there is an indeterminacy in the association of perrotations with a given internal isometric transformation τ. In fact, if the point group for arbitrary γ, denoted $\mathfrak{G}(\gamma)$ (isomorphic with G^P, equation 3.35; we change notation to follow [48]), has order $|\mathfrak{G}(\gamma)|$, then the solutions of the isometry equation 3.63 are $|\mathfrak{G}(\gamma)|$-valued. To see this, we must consider the properties of $\mathfrak{G}(\gamma)$ in more detail.

The covering operation $g \in \mathfrak{G}(\gamma)$ is a rotation or rotation-inversion of the SRMM resulting in a permutation of SRMM vectors associated with identical nuclei (cf. §2.2,§3.1): for any g in $\mathfrak{G}(\gamma)$,

$$g{:}a^\alpha(\gamma)_i \rightarrow R(g)_{ij}a^\alpha(\gamma)_j = \sum a^\beta(\gamma)_i S(g)_{\beta\alpha} \quad , \qquad 3.65$$

so that ([135], equation 2.19'')

$$a^\alpha(\gamma)_i = \sum_\beta a^\beta(\gamma)_j R(g)_{ji} S(g)_{\beta\alpha} \quad . \qquad 3.66$$

Hence, if the perrotation $R(\tau)\otimes S(\tau)$ is a solution of the isometry equation 3.63 associated with the internal isometric transformation τ, then so is $R(g)R(\tau)\otimes S(g)S(\tau)$ for any element g in $\mathfrak{G}(\gamma)$.

In order to formulate a definition of NRM symmetry overcoming this indeterminacy, and encompassing both the notion of point symmetry and of internal isometric transformation, Günthard <u>et al.</u> define the <u>complete isometric group</u> $\overline{\mathcal{H}}(\gamma)$ as the abstract group (complex)

$$\overline{\mathcal{H}}(\gamma) \equiv \mathfrak{G}(\gamma)\cdot\overline{\mathcal{F}}(\gamma) \qquad 3.67$$

where, in practice, a faithful realization of $\overline{\mathcal{H}}(\gamma)$ is constructed as a set of perrotations $\Gamma^{(Ncf)}\{\overline{\mathcal{H}}\}$ on the SRMM configuration $\{a^\alpha(\gamma)_i\}$ ([48]§3.3,2.3).

The question immediately arises: what is the internal structure (if any) of $\overline{\mathcal{H}}(\gamma)$? In [135], the following theorem is proposed: the (abstract) complete isometric group $\overline{\mathcal{H}}(\gamma)$ is a semi-direct product of the point group $\mathcal{G}(\gamma)$ and the internal isometric group $\overline{\overline{\mathcal{F}}}(\gamma)$, where $\mathcal{G}(\gamma)$ is the invariant subgroup.

To prove this theorem, it is first necessary to show that $\mathcal{G}(\gamma)$ is invariant under the internal isometric group $\overline{\mathcal{F}}$, and hence is an invariant subgroup of $\overline{\mathcal{H}}$ (cf. Appendix 3). This result follows directly from the definition of the point group, since the global specification of $\mathcal{G}(\gamma)$ as the highest common symmetry over the whole of the parameter space Γ ensures that, for any τ in $\overline{\mathcal{F}}$,

$$\mathcal{G}(\tau(\gamma)) = \mathcal{G}(\gamma) \qquad 3.68$$

so that

$$\mathcal{G} \lhd \overline{\mathcal{H}} \, . \qquad 3.69$$

Given that $\mathcal{G}$ is an invariant subgroup, it must then be shown that the coset generators of the (abstract) complete isometric group with respect to the point group can be chosen to form a group, isomorphic with the internal isometric group. However, in their treatment of this aspect of the theorem, Günthard et al. proceed by a less direct route. Thus, they choose to determine the representations of $\overline{\mathcal{H}}$ on various substrates of interest (molecular model vectors, rotation coordinates, internal parameters etc.) and attempt to demonstrate semi-direct product structure for each case.

The validity of this approach depends crucially upon the assertion ([135] equation 2.35) that all point symmetry groups can be written as the semi-direct product of two covering (point) groups. However, this is not true in general [125]: for example, the cyclic group of order 4n, C_{4n}, although possessing a proper invariant subgroup C_{2n}, cannot be written as the semi-direct product of its invariant subgroup and another subgroup of order 2 (cf. Appendix 4), and our counterexample will exploit this fact.

Consider first the SRMM for the substituted cyclobutadiene $(CXY)_4$ (atomic X,Y) shown in Figure 3.9, which has covering symmetry $\mathcal{G}(\gamma) = C_{2v}$. In this example, there is a single internal parameter, the angle of pucker γ $(0 \leq \gamma < \pi/2)$ and the

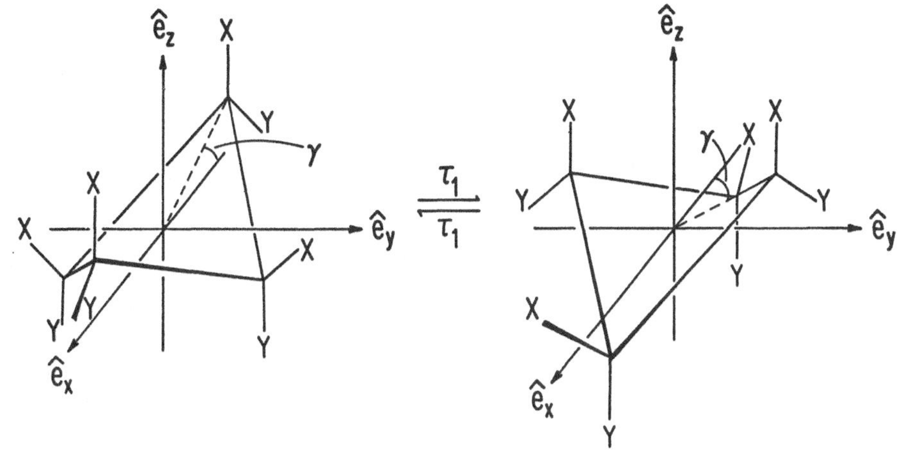

Figure 3.9 The C_{4v} - C_{2v} rotation/ring-puckering SRMM.

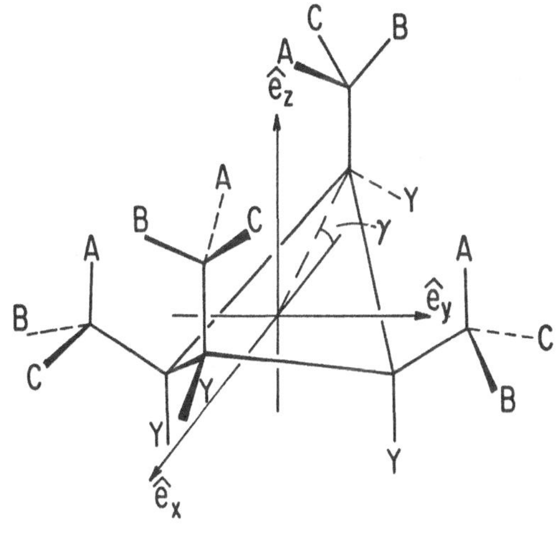

Figure 3.10 The C_4 -C_2 rotation/ring-puckering SRMM.

transformation

$$\tau_1 : \gamma \to -\gamma \qquad \text{(pucker)} \qquad 3.70$$

shown is an isometry. This molecular model is a useful starting point for our discussion, since it obeys the assertions of the general theorem.

The properties of the isometric group for this SRMM are summarized as follows $(\tau_0 : \gamma \to \gamma)$:

$$\overline{\mathcal{F}} = \{\tau_0, \tau_1\} \overset{\text{iso}}{=} \mathcal{V}_2 \qquad 3.71a$$

$$\mathcal{G}(\gamma) = C_{2v} \overset{\text{iso}}{=} \Gamma^{(3)}\{\mathcal{G}\} \qquad 3.71b$$

$$\overline{\mathcal{H}} = C_{2v} \cdot \mathcal{V}_2 \qquad 3.71c$$

$$C_{2v} \triangleleft \overline{\mathcal{H}} \qquad 3.71d$$

$$\Gamma^{(3)}\{\overline{\mathcal{L}}\} \overset{\text{iso}}{=} C_{4v} \qquad 3.71e$$

$$C_{2v} \triangleleft C_{4v} \qquad 3.71f$$

and

$$\mathcal{K} \equiv \Gamma^{(3)}\{\overline{\mathcal{L}}\}/\Gamma^{(3)}\{\mathcal{G}\} = C_{4v}/C_2 \overset{\text{iso}}{=} \mathcal{V}_2 \quad . \qquad 3.71g$$

Thus, the internal isometric group is of order 2, the point group is C_{2v}, and $\overline{\mathcal{H}}$ is of order 8. The set $\Gamma^{(3)}\{\mathcal{G}\}$ is the defining representation of $\mathcal{G}(\gamma)$

$$\Gamma^{(3)}\{\mathcal{G}\} \equiv \{R(g) | g \in \mathcal{G}(\gamma)\} \qquad 3.72$$

isomorphic with C_{2v}, while $\Gamma^{(3)}\{\overline{\mathcal{L}}\}$ is the set of all 3 by 3 orthogonal matrices associated with $\overline{\mathcal{H}}$, and is isomorphic with C_{4v}, which is the effective point symmetry of a puckering 4-ring. The SRMM of Figure 3.9 can therefore be denoted $(CXY)_4\ C_{4v}\text{-}C_{2v}$.

In this case, the effective point group C_{4v} can be written as a semi-direct product of the point group C_{2v} with the reflection group generated by $\hat{\sigma}_d$, the reflection in a dihedral plane

$$C_{4v} = C_{2v} \circledS C_s(\sigma_d) \qquad 3.73$$

i.e., $$\Gamma^{(3)}\{\mathcal{L}\} = \Gamma^{(3)}\{\mathcal{G}\} \circledS \Gamma^{(3)}\{\mathcal{K}\} \qquad 3.74a$$

where $$\Gamma^{(3)}\{\mathcal{K}\} \equiv \{R(\hat{E}),\ R(\hat{\sigma}_d)\} \overset{iso}{\equiv} \mathcal{V}_2 \quad . \qquad 3.74b$$

The essential condition on the structure of the group $\Gamma^{(3)}\{\mathcal{K}\}$ is therefore satisfied, so we can conclude from [135] that representations of $\bar{\mathcal{H}}$ for the $(CXY)_4$ C_{4v}-C_{2v} model exhibit semi-direct product structure, and that the complete isometric group is a semi-direct product

$$(CXY)_4\ C_{4v}\text{-}\ C_{2v} \qquad \bar{\mathcal{H}} = C_{2v} \circledS \mathcal{V}_2 \overset{iso}{\equiv} C_{4v} \quad . \qquad 3.75$$

Note that it is only possible to find the group $\Gamma^{(3)}\{\mathcal{K}\}$ of orthogonal matrices isomorphic with the factor group $\mathcal{K}$ because of the presence of the self-inverse reflection $\hat{\sigma}_d$. There is then a 1:1 correspondence between the elements of $\mathcal{K}$ and $\Gamma^{(3)}\{\mathcal{K}\}$.

With this in mind, we see that in order to exhibit a counter-example to the theorem we must find a model where (for example) $\mathcal{K} \overset{iso}{\equiv} \mathcal{V}_2 \overset{iso}{\equiv} \bar{\bar{\mathcal{F}}}$, but $\Gamma^{(3)}\{\mathcal{L}\}$ cannot be generated from $\Gamma^{(3)}\{\mathcal{G}\}$ by a self-inverse element. This can be done by removing the reflection symmetries from the previous example; the required substitution of chiral groups X,Y produces the C_4-C_2 model shown in Figure 3.10, whose properties are as follows:

$$\bar{\bar{\mathcal{F}}} = \{\tau_0, \tau_1\} \overset{iso}{\equiv} \mathcal{V}_2 \qquad 3.76a$$

$$\mathcal{G}(\gamma) = C_2 \overset{iso}{\equiv} \Gamma^{(3)}\{\mathcal{G}\} \qquad 3.77b$$

$$\bar{\mathcal{H}} = C_2 \cdot \mathcal{V}_2 \qquad 3.76c$$

$$C_2 \lhd \bar{\mathcal{H}} \qquad 3.76d$$

$$\Gamma^{(3)}\{\mathcal{L}\} \overset{iso}{\equiv} C_4 \qquad 3.76e$$

$$C_2 \lhd C_4 \qquad 3.76f$$

and $$\mathcal{K} = C_4/C_2 \overset{iso}{\equiv} \mathcal{V}_2 \qquad 3.76g$$

The internal isometric group $\bar{\bar{\mathcal{F}}}$ is the same as before, being generated by the puckering transformation τ_1 (the C-A bonds are constrained always to lie, for example, in the vertical plane). Here, the effective point symmetry is C_4, the cyclic group of order 4, and $\overline{\overline{\mathcal{H}}}$ is also of order 4. The point group C_2 is an invariant subgroup of C_4. However, in contrast to the previous example, it is not possible to construct C_4 as a semi-direct product of C_2 and a point group of order 2, so that the proof of the second part of the theorem attempted in [135] fails at this point.

We therefore have a counterexample to the theorem of Günthard et al.: for the C_4-C_2 molecular model shown in Figure 3.10, the complete isometric group $\overline{\overline{\mathcal{H}}}$ is not a semi-direct product. The non-existence of a general theorem on the structure of the isometric group is in accord with the discussion of §3.2 (cf. [49]), in that semi-direct product structure does not necessarily emerge from a general analysis of the NRM symmetry group.

Having established a particular counterexample, it is appropriate to inquire more deeply into the general reasons for the failure of this aspect of the proof of the theorem. It appears that the definition of the group $\Gamma^{(3)}\{\mathcal{K}\}$ is of central importance here. In their work, Günthard et al. define the group $\Gamma^{(3)}\{\mathcal{K}\}$ as follows ([135], equation 2.12)

$$\Gamma^{(3)}\{\mathcal{K}\} \equiv \{R(\tau) \mid \tau \in \overline{\overline{\mathcal{F}}}(\gamma)\} \qquad 3.77$$

i.e., the set of all different rotational parts $R(\tau)$ of the representation $\Gamma^{(Ncf)}\{\overline{\overline{\mathcal{F}}}\}$ (3.64), and assert that there is in general a homomorphism $\hat{\eta}$ from $\Gamma^{(Ncf)}\{\overline{\overline{\mathcal{F}}}\}$ onto the matrix group $\Gamma^{(3)}\{\mathcal{K}\}$ ([135], equation 2.13)

$$\hat{\eta}: \Gamma^{(Ncf)}\{\overline{\overline{\mathcal{F}}}\} \xrightarrow{\text{ho}} \Gamma^{(3)}\{\mathcal{K}\} \qquad 3.78a$$

such that

$$\Gamma^{(Ncf)}\{\overline{\overline{\mathcal{F}}}\}/(\text{Ker. } \hat{\eta}) \overset{\text{iso}}{=} \Gamma^{(3)}\{\mathcal{K}\} \quad . \qquad 3.78b$$

However, we would argue that this actually begs the question of the possible semi-factorization of $\Gamma^{(3)}\{\overline{\overline{\mathcal{L}}}\}$ (and hence $\overline{\overline{\mathcal{H}}}$) as a semi-direct product with invariant subgroup $\mathcal{G}$. Thus, in general only the following entities are well defined: the matrix group $\Gamma^{(3)}\{\overline{\overline{\mathcal{L}}}\}$; the matrix group $\Gamma^{(3)}\{\mathcal{G}\}$; the factor group $\mathcal{K}$

$$\mathcal{K} \equiv \Gamma^{(3)}\{\overline{\mathcal{L}}\}/\Gamma^{(3)}\{\mathcal{G}\} \quad . \tag{3.79}$$

It follows that <u>if</u> a matrix group $\Gamma^{(3)}\{\mathcal{K}\}$, isomorphic with $\mathcal{K}$, exists such that $\Gamma^{(3)}\{\overline{\mathcal{L}}\}$ can be written as a semi-direct product

$$\Gamma^{(3)}\{\overline{\mathcal{L}}\} = \Gamma^{(3)}\{\mathcal{G}\} \circledS \Gamma^{(3)}\{\mathcal{K}\} \tag{3.80}$$

<u>then</u> it is true that there is a homomorphism from $\Gamma^{(Ncf)}\{\overline{\overline{\mathcal{F}}}\}$ (the kernel of which is the intrinsic group; cf. 3.32) onto $\Gamma^{(3)}\{\mathcal{K}\}$, and that $\overline{\overline{\mathcal{H}}}$ is also a semi-direct product. We therefore have a necessary condition for the validity of the theorem. An exhaustive examination of the semi-factorizability condition 3.80 is given by Günthard <u>et al.</u> in [185] (see also Appendix 4; [125]).

The problem with the discussion in [135] is that, in the absence of the relation 3.80 between $\Gamma^{(3)}\{\overline{\mathcal{L}}\}$, $\Gamma^{(3)}\{\mathcal{G}\}$ and $\mathcal{K}$, it is not necessarily possible to find the many:one correspondence between elements of $\overline{\overline{\mathcal{F}}}$ and $\mathcal{K}$ implied by the definitions 3.77 and 3.78. This is confirmed by the existence of the C_4-C_2 counterexample.

Finally, we consider the symmetry group $\mathbf{H}$ of the C_4-C_2 molecular model. This is the set of 4 elements

$$\mathbf{H} = \{(\hat{E},\tau_0),(\hat{C}_{4z},\tau_1),(\hat{C}_{2z},\tau_0),(\hat{C}^3_{4z},\tau_1)\} \tag{3.81a}$$

$$\underset{=}{\text{iso}} \quad C_4 \tag{3.81b}$$

isomorphic with the cyclic group of order 4 (as is the PI group for the problem; cf. 3.76c). The point group $\mathbf{G}^P$ is the set of 2 elements

$$\mathbf{G}^P = \{(\hat{E},\tau_0),(\hat{C}_{2z},\tau_0)\} \tag{3.82a}$$

$$\underset{=}{\text{iso}} \quad C_2 \tag{3.82b}$$

isomorphic with the cyclic group of order 2 (cf. 3.76b). It follows from the multiplication rule 3.10, which is valid here, that $\mathbf{G}^P$ is an invariant subgroup of $\mathbf{H}$ (3.76d)

$$G^P \lhd H \quad , \qquad 3.83$$

and that the factor group is the abstract group of order 2 (3.76g)

$$H/G^P \overset{\text{iso}}{=} \sim V_2 \quad . \qquad 3.84$$

Decomposing H into cosets with respect to the invariant subgroup G^P

$$H = \{(\hat{E},\tau_0),(\hat{C}_{2z},\tau_0)\} \cup \{(\hat{C}_{4z},\tau_1),(\hat{C}^3_{4z},\tau_1)\} \qquad 3.85$$

it can easily be seen that it is not possible to find coset generators that form a group of order 2, and thereby construct H as a semi-direct product. Thus, for the C_4-C_2 molecular model of Figure 3.10, H is not a semi-direct product of G^P and another group of order 2, and this corresponds to our conclusion concerning the isometric group.

We would stress, however, that for the symmetry group of the molecular model this result is manifest directly in the coset decomposition 3.85. In order to specify an element of H, we must give both an (internal isometric) transformation τ and a rotation ρ; using the multiplication rule 3.10, the elements of H can then be manipulated directly. It is this feature that gives us a firm hold on the structure of the NRM symmetry group, and which leads to a relatively simple formulation of the problem.

Chapter 4. Nonrigid Molecule Symmetry Groups

Bringing together the formalism of the previous chapter and the semi-direct product theory described in Appendix 3, it is possible to embark upon an exploration of the structure of NRM symmetry groups. The treatment given in this chapter is certainly neither exhaustive nor systematic, but is intended to illustrate the application of our approach to some representative NRMs and, we hope, establish its practical significance. We shall in fact mainly be concerned with NRMs exhibiting some form of internal rotation (cf. [123, 125, 138]); nevertheless, §4.5 treats the symmetry group of the pseudorotating XY_4Z molecule [68], while two ring-puckering models have been discussed in §3.4.

Before turning to the details of the NRM symmetry groups, we would emphasize three general points:

First, if we are able to write the symmetry group of the SRMM in semi-direct product form such as 3.43 or 3.55, then the resulting specification of the NRM group is very much more precise and useful than the description "G_n" [96], where n is the order of the group (a large number). This idea seems first to have been taken up by Altmann [124], although it was perhaps implicit in the work of McIntosh [146, 147]. In a wider context, the construction of a large (possibly infinite) group in terms of a small number of generators is a well established approach to the investigation of group structure ([152, 153]; cf. also [112, 187]). On the basis of semi-direct product structure, we are for example able in this chapter to elucidate the general properties of the symmetry groups for the coaxial rotor NRMs XY_n-XY_n (n odd <u>and</u> even).

Second, when the NRM symmetry group is a semi-direct product, its class structure and character table can be calculated using systematic, as opposed to *ad hoc* [136], methods (Appendix 3). (Even when H is not actually a semi-direct product, it is in general an *extension* of the direct product $G^P \otimes G^I$ [187].)

Third, and most important from the point of view of potential applications to the description of nuclear vibrations in NRMs [126, 154], the semi-direct product nomenclature for NRM group irreducible representations incorporates a great deal of essential information concerning the correlation of NRM states with the various point symmetries occuring over the whole parameter range. Thus, given a knowledge of the orbits of the invariant subgroup under $\mathbf{K}$ or $\tilde{\mathbf{K}}$, rigid/nonrigid correlation diagrams can be constructed virtually by inspection, demonstrating once again the advantages to be gained from this formulation of the theory of NRM symmetry.

4.1 The XY_3-invertor

It has already been established in Chapter 3 that the symmetry group of the XY_3-invertor model shown in Figure 3.1 is the direct product (equation 3.46)

$$\mathbf{H} = C^P_{3v} \otimes C_h \overset{\text{iso}}{\cong} D_{3h} \qquad 4.1$$

so that in this case the IRs of $\mathbf{H}$ are very simply related to those of the invariant point subgroup C^P_{3v}, and we have the rigid/nonrigid correlation diagram given in Figure 4.1 (cf. [82], Figures 3 and 4; [96], Figures 19 and 20; the semi-direct product notation for the point group IRs is developed in Appendix 4):

Figure 4.1. Rigid/nonrigid correlation diagrams for the XY_3-invertor model

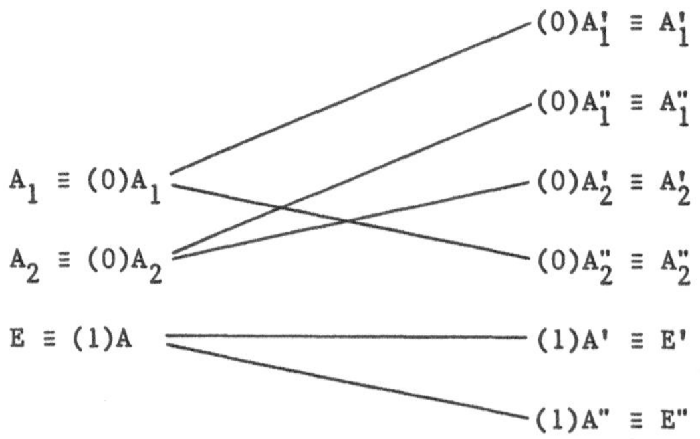

$C_{3v} = C_3 \circledwedge C_s$ $\qquad\qquad$ $D_{3h} = D_3 \otimes C_h$

This diagram may be taken to describe the inversion splittings of the levels in a nominally C_{3v} molecule such as NH_3; the correlation diagrams derived below for more complicated NRMs therefore represent generalized 'contortion' splittings.

4.2 Coaxial rotors with non-identical ends: XY_n-$X'Y'_{n'}$

(a) CH_3-NO_2

This molecular model was described in Chapter 3 (see Figure 3.2), where it was shown that the symmetry group of the SRMM is the semi-direct product

$$G_{12} = \mathbf{C}_3^I \circledS \mathbf{C}_{2v} , \qquad 4.2$$

where

$$\mathbf{C}_3^I = \{h_I^{\mu_I} \equiv (\mu_I) | \mu_I = 0,1,2\} \qquad 4.3a$$

$$h_I \equiv (\hat{E}, \gamma \to \gamma + 2\pi/3) \qquad 4.3b$$

$$\mathbf{C}_{2v} \equiv \{E, C_{2v}, \sigma, \sigma' = \sigma \cdot C_{2z}\} \qquad 4.3c$$

$$C_{2z} \equiv (\hat{C}_{2z}, \gamma + \pi) \qquad 4.3d$$

$$\sigma \equiv (\hat{\sigma}_{yz}, -\gamma) . \qquad 4.3e$$

The intrinsic operation h_I corresponds to rotation of the methyl group through $2\pi/3$ about the C-N bond, a symmetry operation which clearly involves no overall rotation of the molecular model. Permutations of identical nuclei induced by operations of the symmetry group H are easily determined by reference to Figure 3.4, which shows the action of elements of G_{12} upon the molecular model (cf. equation 3.7).

The class structure of the semi-direct product G_{12} is

$[E \,|(0)]$

$[E \,|(1)]$, $[E \,|(2)]$

$[\sigma \,|(0)]$, $[\sigma \,|(1)]$, $[\sigma \,|(2)]$

$[C_{2z} \,|(0)]$

$[C_{2z} \,|(1)]$, $[C_{2z} \,|(2)]$

$[\sigma' \,|(0)]$, $[\sigma' \,|(1)]$, $[\sigma' \,|(2)]$.

The character table for the invariant cyclic subgroup C_3^I is very simple, as is that for the (factor group) $\mathbf{C}_{2v}$ (cf. Appendix 4). Orbits of C_3^I under C_{2v} are (Appendix 3; note that elements and irreducible representations of cyclic groups C_n

are both written in the same form, i.e., (μ): it should always be clear from the context which is meant):

Orbit	Little co-group
$\{(0)\}$	${}_0\mathcal{K} = C_{2v}$
$\{(1),(2)\}$	${}_1\mathcal{K} = C_2 = \{E, C_{2v}\}$

and the character table for G_{12} is given in Table 4.1. Semi-direct product notation is used for the IRs of G_{12}, the labels used by Longuet-Higgins [1] being given for comparison. Although G_{12} does in fact have a simple direct product structure, being isomorphic with D_{3h}, this example provides a straightforward illustration of the general semi-direct product construction of the IRs and character table.

Table 4.1. Character table for the semi-direct product $C_3^I \circledS C_{2v}$, symmetry group of the NRM $CH_3 - NO_2$.

$C_3^I \circledS C_{2v}$	1	2	3	1	2	3
	$[E\|0]$	$[E\|1]$	$[\sigma_v\|0]$	$[C_{2z}\|0]$	$[C_{2z}\|1]$	$[\sigma'\|0]$
$A_1' = (0)A_1$	1	1	1	1	1	1
$A_2' = (0)A_2$	1	1	-1	1	1	-1
$A_1'' = (0)B_1$	1	1	1	-1	-1	-1
$A_2'' = (0)B_2$	1	1	-1	-1	-1	1
$E' = (1)A$	2	-1	0	2	-1	0
$E'' = (1)B$	2	-1	0	-2	1	0

The correlation diagram of Figure 4.2 relates the IRs of the local symmetry groups for the 'C_{2v} frame' and 'C_{3v}-top' fragments of the molecule to those of the complete NRM group.

Figure 4.2. Rigid/Nonrigid correlation diagram for the NRM CH_3-NO_2

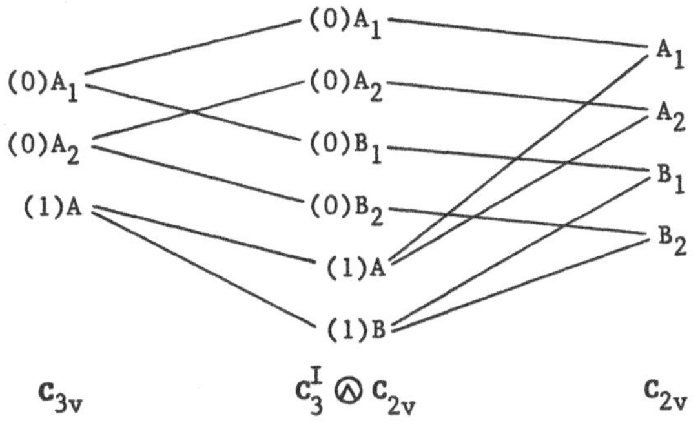

(b) CH_3 - SiH_3

This molecular model is shown in Figure 4.3, where the heavier -SiH_3 group is taken to be the frame while the methyl group is the top.

The symmetry group of the SRMM is the semi-direct product

$$\mathbf{H} = (\mathbf{C}_3^I \otimes \mathbf{C}_3^P) \circledS \mathbf{C}_s \qquad 4.4$$

of order 18, where both the intrinsic group and the point groups are cyclic groups of order 3:

$$\mathbf{C}_3^I = \{h_I^{\mu_I} \equiv (\mu_I) | \mu_I = 0,\ 1,\ 2\} \qquad 4.5a$$

$$h_I \equiv (\hat{E},\ \gamma \rightarrow \gamma + 2\pi/3) \qquad 4.5b$$

$$\mathbf{C}_3^P = \{C_3^{\mu_P} | \mu_P = 0,1,2\} \qquad 4.5c$$

$$C_3 \equiv (\hat{C}_3, \gamma) \quad , \qquad 4.5d$$

and the factor group is isomorphic with a reflection group of order 2:

$$\mathbf{C}_s = \{E,\sigma\} \equiv \{(\hat{E},\gamma),(\hat{\sigma}_{yz},-\gamma)\} \quad . \qquad 4.6$$

Elements and irreducible representations of the direct product invariant subgroup $\mathbf{C}_3^I \otimes \mathbf{C}_3^P$ are both written in the form

$$(\mu_I)(\mu_P) \qquad \mu_I,\ \mu_P \ \underline{\text{mod}}\ 3 \ . \qquad 4.7$$

The procedures described in Appendix 3 give the class structure for **H**:

[E|(0)(0)]

[E|(1)(0)], [E|(2)(0)]

[E|(0)(1)], [E|(0)(2)]

[E|(1)(1)], [E|(2)(2)]

[E|(1)(2)], [E|(2)(1)]

9 { [σ|(0)(0)], [σ|(1)(0)], [σ|(2)(0)]
[σ|(0)(1)], [σ|(0)(2)], [σ|(1)(1)]
[σ|(1)(2)], [σ|(2)(1)], [σ|(2)(2)] ,

and the orbits of $C_3^I \otimes C_3^P$ under C_s

Orbit	Little Co-group
{(0)(0)}	${}_{00}\mathcal{K} = C_s$
{(0)(1), (0)(2)}	${}_{01}\mathcal{K} = \{E\}$
{(1)(0), (2)(0)}	${}_{10}\mathcal{K} = \{E\}$
{(1)(1), (2)(2)}	${}_{22}\mathcal{K} = \{E\}$
{(1)(2), (2)(1)}	${}_{12}\mathcal{K} = \{E\}$.

The character table of the semi-direct product $(C_3^I \otimes C_3^P) \owedge C_s$ is shown in Table 4.2, in agreement with that given by Bunker for the PI group [155].

Table 4.2. Character table for $(C_3^I \otimes C_3^P) \owedge C_s$, symmetry group of the NRM $CH_3 - SiH_3$.

$(C_3^I \otimes C_3^P) \owedge C_s$	1 [E|(0)(0)]	2 [E|(0)(0)]	2 [E|(0)(1)]	2 [E|(1)(1)]	2 [E|(1)(2)]	9 [σ|(0)(0)]
$(0)(0)A_1$	1	1	1	1	1	1
$(0)(0)A_2$	1	1	1	1	1	-1
(1)(0)A	2	-1	2	-1	-1	0
(0)(1)A	2	2	-1	-1	-1	0
(1)(1)A	2	-1	-1	-1	2	0
(2)(1)A	2	-1	-1	2	-1	0

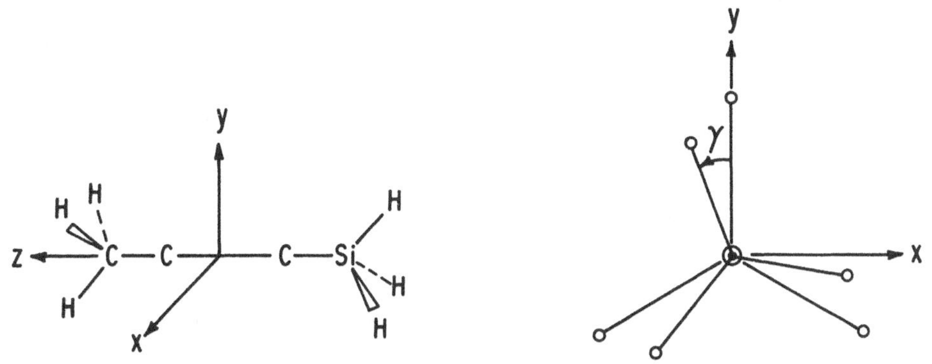

Figure 4.3 The SRMM for CH_3 - SiH_3

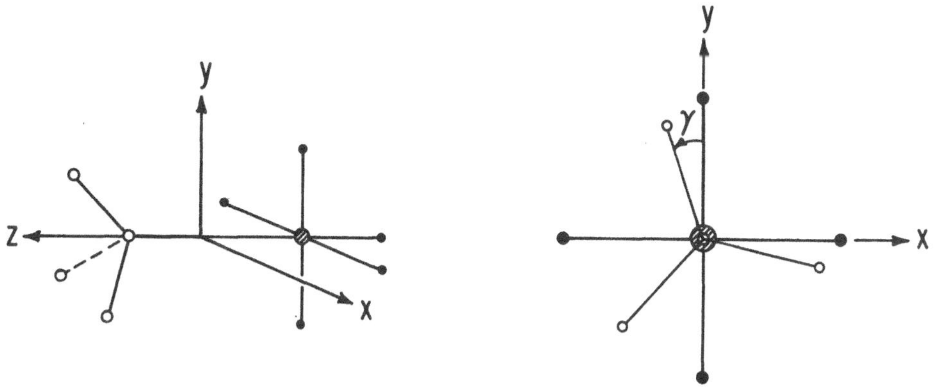

Figure 4.5 The SRMM for CH_3 - SF_5

Figure 4.4 correlates the IRs of the point subgroup C_3^P with those of the NRM group <u>via</u> IRs of the group $C_3^P \rtimes C_s \overset{iso}{\equiv} C_{3v}$, which is a covering group of the molecular model for values of the parameter $\gamma = 0, \pi$ ('fixed points' in the terminology of the isometric theory [48]).

Figure 4.4. Rigid/Nonrigid correlation diagram for the NRM $CH_3 - SiH_3$

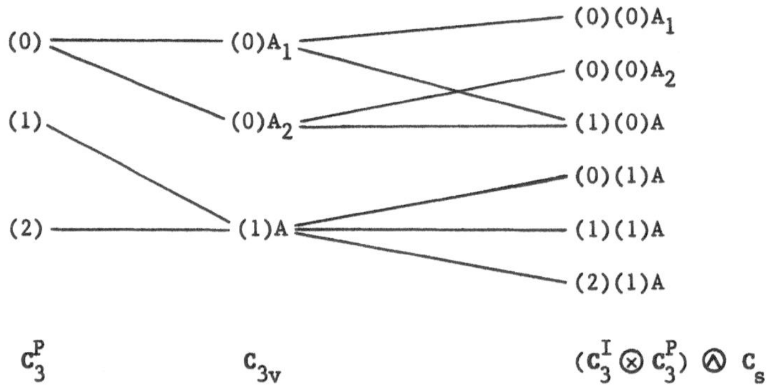

(c) $CH_3 - SF_5$

This molecular model is shown in Figure 4.5; the heavy $-SF_5$ group is taken to be the frame and the $-CH_3$ group is the top.

The symmetry group of the SRMM is a semi-direct product of order 24

$$H = C_3^I \rtimes (C_4 \rtimes C_s) \qquad 4.8a$$

$$\equiv (C_3^I \otimes C_4) \rtimes C_s \, , \qquad 4.8b$$

i.e.,

$$K = C_4 \rtimes C_s \qquad 4.9$$

where the operations generating the groups C_3^I and C_s are identical with those defined above, while

$$C_4 = \{C_{4z}^{\mu};\ \mu = 0,1,2,3\} \qquad 4.10a$$

$$C_{4z} \equiv (\hat{C}_{4z},\ \gamma \rightarrow \gamma + \pi/2) \quad . \qquad 4.10b$$

The group $(C_3^I \otimes C_4) \rtimes C_s$ is in fact isomorphic with D_{12} (cf. Appendix 4), having two classes consisting of one element each, five classes of two elements each and two classes of six elements.

Orbits of $C_3^I \otimes C_4$ under C_s are determined to be:

Orbit	Little Co-group
{(0)(0)}	$_{00}\mathcal{K} = C_s$
{(0)(2)}	$_{02}\mathcal{K} = C_s$
{(0)(1), (0)(3)}	$_{01}\mathcal{K} = \{E\}$
{(1)(0), (2)(0)}	$_{10}\mathcal{K} = \{E\}$
{(1)(1), (2)(3)}	$_{11}\mathcal{K} = \{E\}$
{(1)(2), (2)(2)}	$_{12}\mathcal{K} = \{E\}$
{(1)(3), (2)(1)}	$_{13}\mathcal{K} = \{E\}$

and the character table for $(C_3^I \otimes C_4) \circledS C_s$ is given in Table 4.3.

Table 4.3. Character Table for $(C_3^I \otimes C_4) \circledS C_s$, symmetry group of the NRM $CH_3 - SF_5$

C_s / $C_3^I \otimes C_4$	1 E (0)(0)	2 E (1)(3)	2 E (1)(2)	2 E (0)(1)	2 E (1)(0)	2 E (1)(1)	1 E (0)(2)	6 σ (0)(0)	6 σ (0)(1)
(0)(0).A_1	1	1	1	1	1	1	1	1	1
(0)(0).A_2	1	1	1	1	1	1	1	-1	-1
(0)(2).A_1	1	-1	1	-1	1	-1	1	1	-1
(0)(2).A_2	1	-1	1	-1	1	-1	1	-1	1
(0)(1).A	2	0	-2	0	2	0	-2	0	0
(1)(0).A	2	-1	-1	2	-1	-1	2	0	0
(1)(1).A	2	$\sqrt{3}$	1	0	-1	$\sqrt{3}$	-2	0	0
(1)(2).A	2	1	-1	-2	-1	1	2	0	0
(1)(3).A	2	$-\sqrt{3}$	1	0	-1	$\sqrt{3}$	-2	0	0

Figure 4.6 correlates the IRs of the NRM group **H** with those of the invariant subgroup C_3^I and fixed-point symmetry C_s ($\gamma = 0$).

Figure 4.6. Correlation of the IRs of $(C_3^I \otimes C_4) \circledS C_s$ with those of C_3^I and C_s.

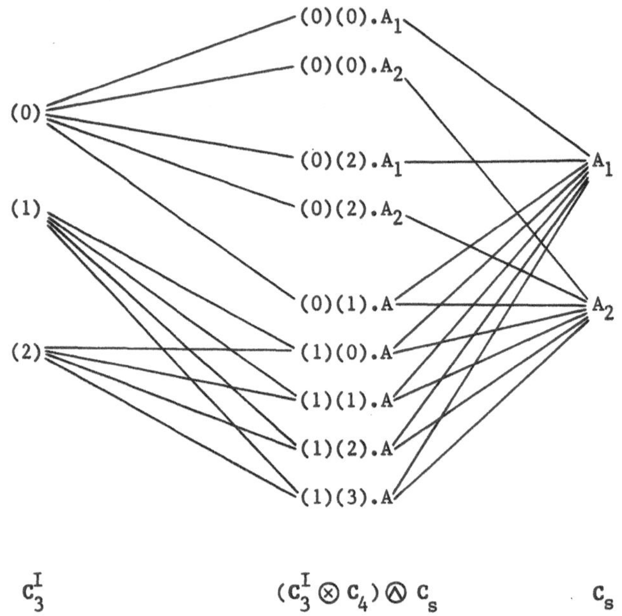

4.3 Coaxial rotors with identical ends: $XY_n - XY_n$

We now come to the problem of the symmetry groups for coaxial rotor models $XY_n - XY_n$ having identical ends.

Several interesting new features arise here. Thus, when the molecular models are parametrized as in Figures 4.7 and 4.8, so that a 'top' end is not distinguished from a 'frame' end, the symmetry operation denoted C_{2z},

$$C_{2z} \equiv (\hat{C}_{2z}, \gamma \rightarrow \gamma + \pi) \quad , \qquad 4.11$$

is a <u>primitive period transformation</u> (PPT): that is, it induces the identity permutation of SRMM vectors (cf. §3.1). This means that the NRM symmetry groups obtained here correspond directly to 'double' PI groups [96], the PI groups themselves (or the H^{π}) being homomorphic images of the SRMM groups H.

When n is odd, it is possible to write H as a direct product involving the group of order 2 generated by the PPT C_{2z}, so that reduction from the double group onto the PI group and conversely extension of the PI group to the double group is trivial. In addition, the PI group can itself be written as a direct product, so that practical construction of the character table for the symmetry group of the coaxial rotor is relatively easy when n is odd [1, 156, 157]. Nevertheless, we shall reexamine these groups here from our general point of view, since it provides information on the correlation of NRM states with those for various point symmetries.

When n is even, the position is apparently more complicated (as noted by Bunker [156]). First of all, there is no convenient direct-product factorization of H as a semi-direct product with invariant subgroup $G^I \otimes G^P$; rather, it is necessary to introduce an extended intrinsic group $\tilde{G}^I$ in order to semi-factorize H (cf. §3.2). Furthermore, the PPT C_{2z} straddles both $\tilde{G}^I$ and G^P and cannot be factored away, thereby complicating the mapping from H onto the PI group.

It is here, however, that the utility of our approach can be appreciated. Thus, it is in principle possible to deal with all the above problems with little more effort than that required when n is odd; in practice, the only difficulty is that we have to handle genuine semi-direct products rather than direct product groups. In this way, we have been able to derive two new NRM group character tables: those for $n = 4$ (G^+_{64}) and $n = 6$ (G^+_{144}) [189]. The character table for the single PI group G_{64} of the $XY_4 - XY_4$ rotor has previously been given by Bunker [96].

(i) n odd.

We begin with the simple case where n is odd. The molecular models for $n = 1,3,5$ are shown in Figure 4.7; representative molecules are H_2O_2 $(n = 1)$ [96], $CH_3 - CH_3$ or $CH_3 - C \equiv C - CH_3$ $(n = 3)$ [53], and ferrocene $(n = 5)$ [157].

In general, for any odd n, we have an intrinsic group G^I of order n isomorphic with the cyclic group C_n

$$G^I = C^I_n \equiv \{h_I^{\mu_I} \mid \mu_I = 0,1,\ldots,n-1\} \qquad 4.12a$$

$$h_I \equiv (\hat{E},\ \gamma \to \gamma + 2\pi/n) \quad , \qquad 4.12b$$

corresponding to n-fold internal rotations, and a point group G^P of order 2n which is the dihedral group D_n, the covering group for arbitrary γ,

$$G^P = D^P_n \equiv C^P_n \circledS C_2(y) \ , \qquad 4.13a$$

$$C^P_n \equiv \{C_{nz}^{\mu_P} \mid \mu_P = 0,1,\ldots,n-1\} \qquad 4.13b$$

$$C_{nz} \equiv (\hat{C}_{nz},\gamma) \qquad 4.13c$$

$$C_2(y) = \{E,C_{2y}\} \equiv \{(\hat{E},\gamma),\ (\hat{C}_{2y},\gamma)\} \ . \qquad 4.13d$$

The symmetry group of the SRMM is a semi-direct product of order $8n^2$

$$H = (C^I_n \otimes D^P_n) \circledS C_{2h} \equiv G^+_{4n^2} \quad , \qquad 4.14$$

where the group C_{2h} isomorphic with the factor group of order 4 is a direct product

$$C_{2h} = C_i \otimes C_2(z) \qquad 4.15a$$

with

$$C_i = \{E,i\} \equiv \{(\hat{E},\gamma),\ (\hat{i},\pi-\gamma)\} \qquad 4.15b$$

and

$$C_{2z} = \{E,C_{2z}\} \equiv \{(\hat{E},\gamma),(\hat{C}_{2z},\ \gamma + \pi)\} \ . \qquad 4.15c$$

Note that the element $C_{2z} = (\hat{C}_{2z},\ \gamma + \pi)$ is the PPT, i.e., increasing the internal

a)

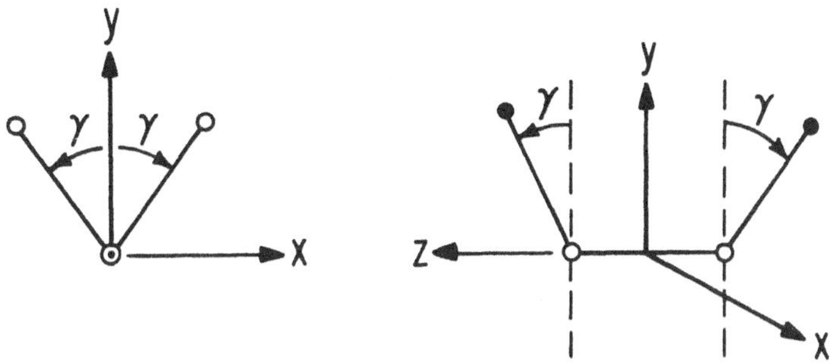

b)

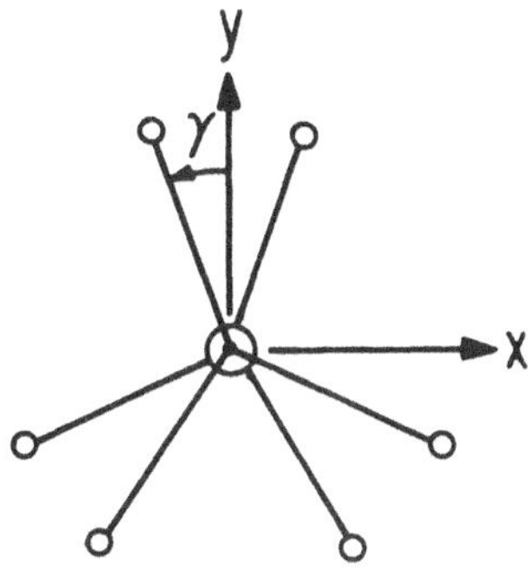

c)

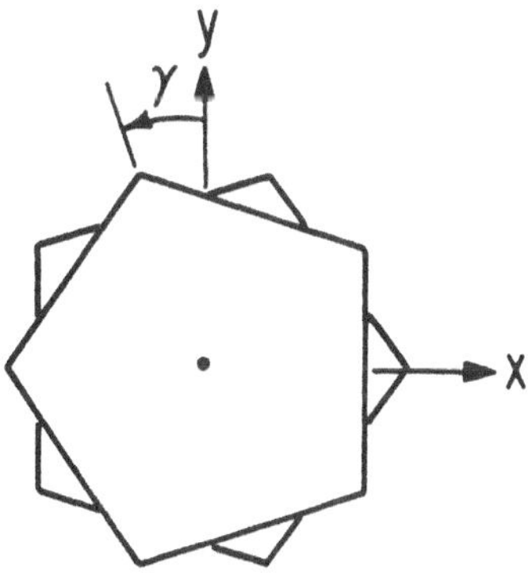

Figure 4.7 SRMMs for coaxial rotors $XY_n - XY_n$ with n odd

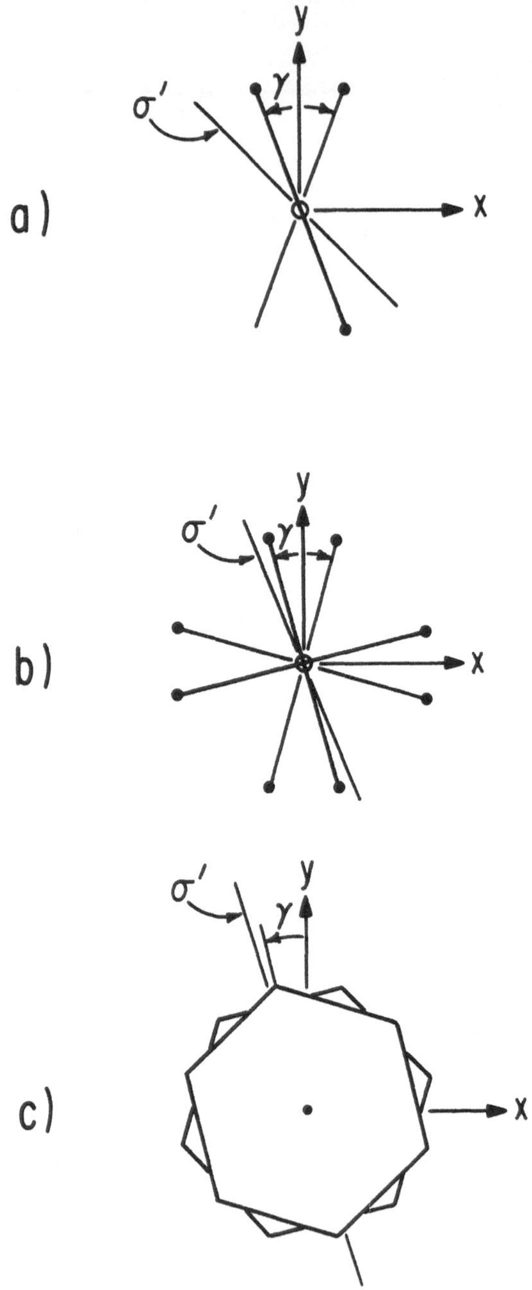

Figure 4.8 SRMMs for coaxial rotors XY_n - XY_n with n even.

parameter γ by π and rotating the model through π about the z-axis leads to the identity permutation of nuclei. A general element of **H** is therefore represented by a product of generating elements

$$h = h_K \cdot h_I^{\mu_I} \cdot h_P \qquad 4.16$$

where h_K is an element of $\mathbf{C}_{2h}$ and h_P is an element of D_n^P.

The symmetry group of the SRMM can be written in direct product form

$$\mathbf{H} = ((\mathbf{C}_n^I \circledS \mathbf{C}_i) \otimes \mathbf{D}_n^P) \otimes \mathbf{C}_2(z) \qquad 4.17a$$

$$\stackrel{iso}{=} (\mathbf{D}_n \otimes \mathbf{D}_n^P) \otimes \mathbf{C}_2(z) \quad , \qquad 4.17b$$

so that factoring off the group $C_2(z)$ generated by the PPT we see that the PI group $\mathbf{H}^\pi$ is isomorphic with the direct product of two D_n groups:

$$\mathbf{H} \xrightarrow{ho.} \mathbf{H}^\pi = \mathbf{D}_n \otimes \mathbf{D}_n^P \qquad 4.18$$

as originally found by Longuet-Higgins [1] and Bunker [156]. The total number of classes of $\mathbf{H} = (\mathbf{C}_n^I \otimes \mathbf{D}_n^P) \circledS \mathbf{C}_{2h}$ is therefore $2 \times [(n+3)/2]^2 = (n+3)^2/2$ (cf. Appendix 4).

We now consider some specific examples.

a) $n = 1$: XY - XY

The molecular model for the coaxial rotor XY - XY is shown in Figure 4.7a. The symmetry group is of order 8, and is in this case a direct product

$$\mathbf{H} = \mathbf{C}_2(y) \otimes \mathbf{C}_{2h} \qquad 4.19a$$

$$\equiv \text{'} G_4^+ \text{'} \qquad 4.19b$$

where the component groups have been defined above. The character table for **H** is given in Table 4.4, where the IRs of the group $\mathbf{C}_{2h}$ are denoted A_{1g}, A_{1u} , A_{2g} and A_{2u}

Table 4.4. Character table for $C_2(y) \otimes C_{2h}$, symmetry group of the NRM XY - XY

$C_2(y) \otimes C_{2h}$	E	C_{2y}	i	iC_{2y}	C_{2z}	$C_{2z}C_{2y}$	σ_{xy}	$\sigma_{xy}C_{2y}$
$A.A_{1g}$	1	1	1	1	1	1	1	1
$A.A_{1u}$	1	1	-1	-1	1	1	-1	-1
$A.A_{2g}$	1	1	1	1	-1	-1	-1	-1
$A.A_{2u}$	1	1	-1	-1	-1	-1	1	1
$B.A_{1g}$	1	-1	1	-1	1	-1	1	-1
$B.A_{1u}$	1	-1	-1	1	1	-1	-1	1
$B.A_{2g}$	1	-1	1	-1	-1	1	-1	1
$B.A_{2u}$	1	-1	-1	1	-1	1	1	-1

Transformation properties of NRM states under the fixed-point covering symmetries $C_2(y) \otimes C_s(\sigma_{yz})$ $(\gamma = 0)$ and $C_2(y) \otimes C_s(\sigma_{xz})$ $(\gamma = \pi/2)$ are easily derived from Table 4.4 by inspection. We note also that there are <u>three</u> PPTs for the special case $n = 1$,

$$C_{2z} = (\hat{C}_{2z}, \gamma+\pi), \; \sigma_{yz} = (\hat{\sigma}_{yz}, -\gamma), \; \sigma_{xZ} = (\hat{\sigma}_{xz}, \pi-\gamma), \qquad 4.20$$

which together with the identity generate a group isomorphic with C_{2v}.

Apart from being appropriate for molecules such as H_2O_2 [96], the group G_4^+ is also the symmetry group of NRMs of the type $CXY_2 - C \equiv C - CXY_2$ [127].

b) $n = 3$: $XY_3 - XY_3$

The molecular model for $XY_3 - XY_3$ is shown in Figure 4.7b, and has the symmetry group

$$H = (C_3^I \otimes D_3^P) \circledwedge C_{2h} \equiv {}'G_{36}^+{}' \qquad 4.21$$

of order 72.

Orbits of $C_3^I \otimes D_3^P$ under C_{2h} are:

Orbit	Little Co-group
$\{(0).(0)A_1\}$	C_{2h}
$\{(0).(0)A_2\}$	C_{2h}
$\{(0).(1)A\}$	C_{2h}
$\{(1).(0)A_1,\ (2).(0)A_1\}$	$C_2(z)$
$\{(1).(0)A_2,\ (2).(0)A_2\}$	$C_2(z)$
$\{(1).(1)A,\ (2).(1)A\}$	$C_2(z)$

where it should be noted that the orbits contain single IRs of D_3^P only, since the elements of D_3^P and C_{2h} mutually commute.

The character table of the semi-direct product $(C_3^I \otimes D_3^P) \circledS C_{2h}$ is identical with that given by Hougen [57] for the double PI group of the dimethylacetylene molecule. Correlation with the homomorphic image G_{36} is immediate, since G_{36}^+ is a direct product (cf. 4.7b).

For the purpose of the correlating the IRs of the NRM group with those of the fixed-point covering symmetries D_{3h} ($\gamma = \pi/2$) it is very useful to consider a subgroup of H of order 24 (compare [53], p.721), which is the direct product $D_3^P \otimes C_{2h}$:

$$H \supset D_3^P \circledS C_{2h} \equiv D_3^P \otimes C_{2h}. \qquad 4.22$$

The IRs of $D_3^P \otimes C_{2h}$ are correlated with those of the point groups D_{3h} and D_{3d} in Figure 4.9. Since the IRs of the semidirect product $H = (C_3^I \otimes D_3^P) \circledS C_{2h}$ are readily correlated with those of its subgroup $D_3^P \otimes C_{2h}$, it can be seen how the semi-direct product formulation of NRM symmetry group facilitates correlation with fixed-point symmetry groups. Such correlations are important in the theory of nuclear vibrations in NRMs.

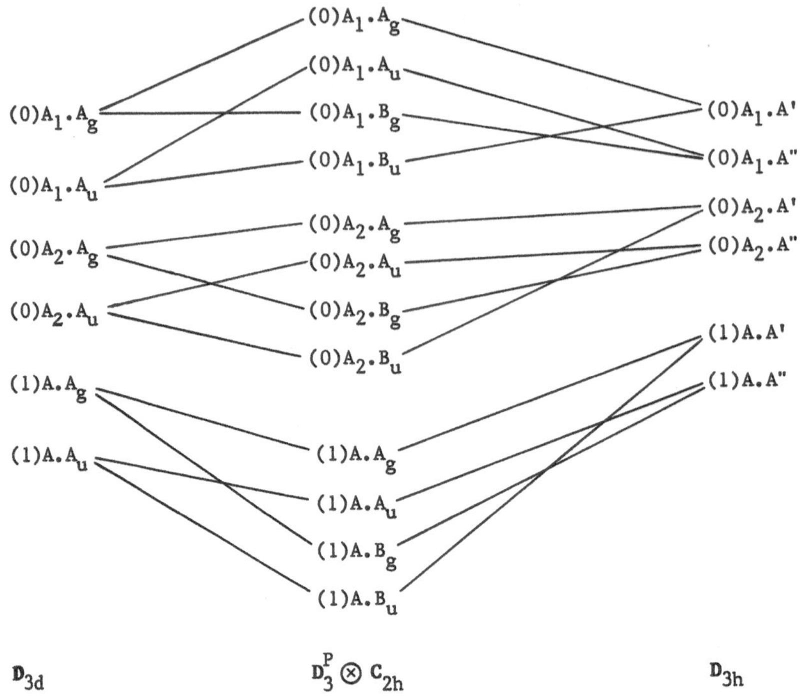

Figure 4.9. Correlation of the IRs of $D_3^P \otimes C_{2h}$ with those of D_{3d} and D_{23h}

c) $n = 5$: $XY_5 - XY_5$

The molecular model for $n = 5$ describing the sandwich compound ferrocene is shown in Figure 4.7c. It has a symmetry group of order 200 which is the semi-direct product

$$H = (C_5^I \otimes D_5^P) \circledS C_{2h} = 'G_{100}^{+}' \quad . \qquad 4.23$$

The character table for the single PI group G_{100}, isomorphic with the direct product $D_5 \otimes D_5$, has been given by Bunker [157].

Some general relations for NRM groups associated with coaxial rotors $XY_n - XY_n$ and $XY_n - X'Y'_n$ (n odd) are shown in Figure 4.20

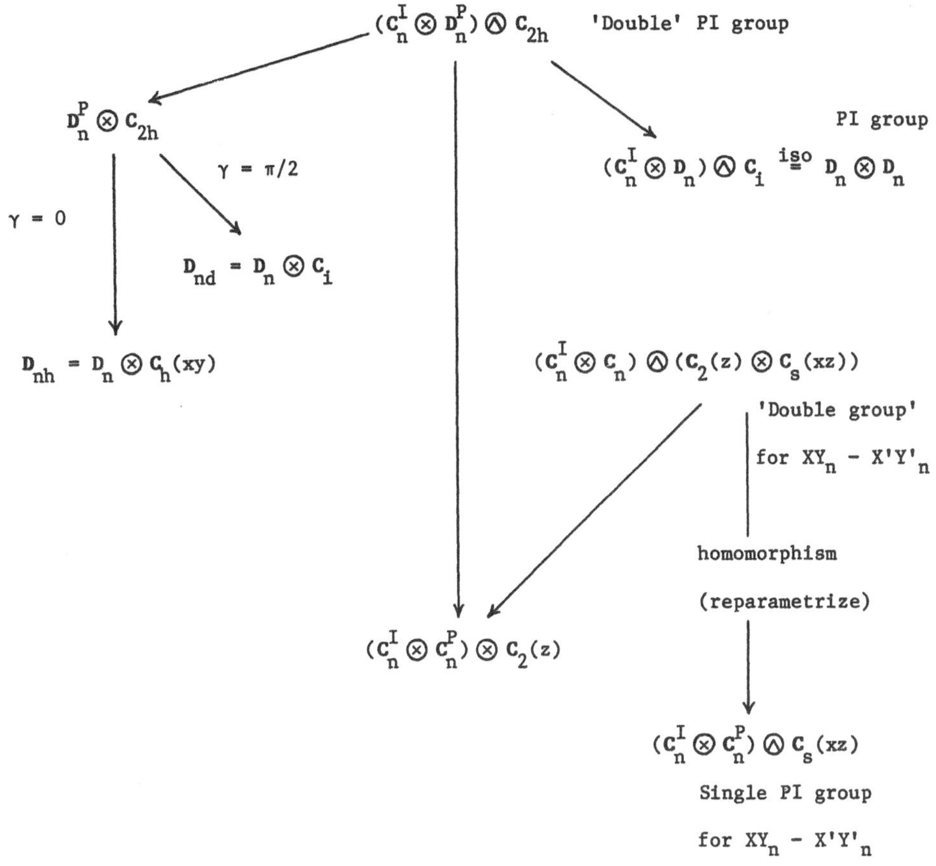

Figure 4.10. Some general relations for the NRM groups associated with rotors XY_n-XY_n and XY_n-$X'Y'_n$ (n odd)

ii) n even

We now consider the more complicated case where n is even. SRMMs for n = 2,4,6 are shown in Figure 4.8; representative molecules are $BF_2 - BF_2$ (n = 2) [128], $SF_5 - SF_5$ (n = 4), and dibenzene chromium (n = 6).

For arbitrary even n the intrinsic group G^I is the group of order n defined in equation 4.12, isomorphic with C_n:

$$G^I = C_n^I \quad , \qquad 4.24$$

and the point group G^P is the group of order 2n defined in equation 4.13, isomorphic with D_n:

$$G^P = D_n^P \ . \tag{4.15}$$

The symmetry group of the SRMM is of order $8n^2$, and has the semi-direct product structure

$$H = (C_{ni}^I \otimes D_n^P) \circledS C_{s'} = {}'G^+_{4n^2}{}' \tag{4.26}$$

where C_{ni}^I is the <u>extended intrinsic group</u> $\tilde{G}^I$ isomorphic with C_{ni}:

$$\tilde{G}^I = C_{ni}^I = C_n^I \otimes C_i \tag{4.27a}$$

$$C_i = \{(\hat{E},\gamma), (\hat{i}, -\gamma)\} \ , \tag{4.27b}$$

while $C_{s'}$ is a group of order 2:

$$C_{s'} = \{E,\sigma'\} = \{(\hat{E},\gamma), (\hat{\sigma}',\pi/n - \gamma)\} \ , \tag{4.28}$$

with the reflection plane σ' lying at an angle $\pi/2n$ to the yz plane (as shown in Figure 4.8, cf. also Appendix 4, Figure A4.2).

We note that for n even it is not possible to semi-factorize H in the form given in 4.17 for odd n; this is due to the presence of a $\hat{C}_{2z}$ rotation operation in both the groups D_n^P (n even) and C_{2h}. In addition, since we have

$$C_{2z} \equiv (\hat{C}_{2z}, \gamma+\pi) = (\hat{C}_{2z},\gamma)(\hat{E},\gamma+\pi) \tag{4.29a}$$

with

$$(\hat{C}_{2z},\gamma) \in D_n^P \quad (\hat{E},\gamma+\pi) \in C_n^I \ , \tag{4.29b}$$

we cannot factor off the PPT operation C_{2z}. It is therefore necessary to find the PI group G_{4n^2} given by the 2:1 homomorphism

$$\left.\begin{array}{l}(\hat{E},\gamma) \\ (\hat{C}_{2z},\gamma+\pi)\end{array}\right\} \longrightarrow \text{Identity} \tag{4.30}$$

i.e.,

$$G^+_{4n^2} = (C_{ni}^I \otimes D_n^P) \circledS C_{s'} \xrightarrow{\text{ho}} G_{4n^2} \ . \tag{4.31}$$

If the character table of $G^{+}_{4n^2}$ can be found using the methods described in Appendix 3, the character table of G_{4n^2} can be formed immediately, since classes of $G^{+}_{4n^2}$ map onto (not necessarily distinct) classes of G_{4n^2} and allowed IRs of the double group are those 'single-valued' IRs having the same character under the operations E and C_{2z}.

Using the methods of Appendix 3 it is possible to determine the general class structure of the NRM group 4.26; the results, which are rather complicated, are given elsewhere [189]. Here we note only that $G^{+}_{4n^2}$ has in general $2(n^2/4 + 3n/2 + 3)$ classes and IRs.

a) n = 2: $XY_2 - XY_2$.

The molecular model for the rotor $XY_2 - XY_2$ is shown in Figure 4.8a; the symmetry group is of order 32:

$$\mathbf{H} = (\mathbf{C}^{I}_{2i} \otimes \mathbf{D}^{P}_{2}) \circledS \mathbf{C}_{s'} = G^{+}_{16} \ . \qquad 4.32$$

Orbits of the invariant subgroup $\mathbf{C}^{I}_{2i} \otimes \mathbf{D}^{P}_{2}$ under $\mathbf{C}_{s'}$ are:

Orbit	Little co-group
$A_g.(0)A_1$	$\mathbf{C}_{s'}$
$A_g.(0)A_2$	$\mathbf{C}_{s'}$
$A_u.(0)A_1$	$\mathbf{C}_{s'}$
$A_u.(0)A_2$	$\mathbf{C}_{s'}$
$A_u.(0)A_2$	$\mathbf{C}_{s'}$
$\{A_g.(1)A_1, A_g.(1)A_2\}$	{E}
$\{A_u.(1)A_1, A_u.(1)A_2\}$	{E}
$\{B_g.(0)A_1, B_u.(0)A_1\}$	{E}
$\{B_g.(0)A_2, B_u.(0)A_2\}$	{E}
$\{B_g.(1)A_1, B_u.(1)A_2\}$	{E}
$\{B_g.(1)A_2, B_u.(1)A_1\}$	{E}

The character table for $(\mathbf{C}^{I}_{2i} \otimes \mathbf{D}^{P}_{2}) \circledS \mathbf{C}_{s'}$ is shown in Table 4.5, and is in agreement with that given by Watson and Merer [128]. (Note that

$$C_2 \equiv (\hat{E}, \gamma+\pi), \quad i \equiv (\hat{i}, -\gamma) \quad \text{in Table 4.5.} \qquad 4.33$$

Table 4.5. Character table for $(C^{I}_{2i} \otimes D^{P}_{2}) \circledwedge C_{s'}$, symmetry group of the NRM XY_2 - XY_2.

	1	1	1	1	2	2	2	2	2	2	4	4	4	4
$C_{s'}$	E	E	E	E	E	E	E	E	E	E	σ'	σ'	σ'	σ'
C^{I}_{2i}	E	E	C_2	C_2	E	C_2	i	i	i	i	E	E	i	i
D^{P}_{2}	E	C_{2z}	E	C_{2z}	C_{2y}	C_{2y}	E	C_{2z}	C_{2y}	C_{2x}	E	C_{2y}	E	C_{2y}
$(A_g\cdot(0)A_1)A''$	1	1	1	1	1	1	1	1	1	1	1	1	1	1
$(A_g\cdot(0)A_1)A''$	1	1	1	1	1	1	1	1	1	1	-1	-1	-1	-1
$(A_g\cdot(0)A_2)A'$	1	1	1	1	-1	-1	1	1	-1	-1	1	-1	1	-1
$(A_g\cdot(0)A_2)A''$	1	1	1	1	-1	-1	1	1	-1	-1	-1	1	-1	1
$(A_u\cdot(0)A_1)A'$	1	1	1	1	1	1	-1	-1	-1	-1	1	1	-1	-1
$(A_u\cdot(0)A_1)A''$	1	1	1	1	1	1	-1	-1	-1	-1	-1	-1	1	1
$(A_u\cdot(0)A_2)A'$	1	1	1	1	-1	-1	-1	-1	1	1	1	-1	-1	1
$(A_u\cdot(0)A_2)A''$	1	1	1	1	-1	-1	-1	-1	1	1	-1	1	1	-1
$(A_g\cdot(1)A_1)A$	2	-2	2	-2	0	0	2	-2	0	0	0	0	0	0
$(A_u\cdot(1)A_1)A$	2	-2	2	-2	0	0	-2	2	0	0	0	0	0	0
$(B_g\cdot(0)A_1)A$	2	2	-2	-2	2	-2	0	0	0	0	0	0	0	0
$(B_g\cdot(0)A_2)A$	2	2	-2	-2	-2	2	0	0	0	0	0	0	0	0
$(B_g\cdot(1)A_1)A$	2	-2	-2	2	0	0	0	0	2	-2	0	0	0	0
$(B_g\cdot(1)A_2)A$	2	-2	-2	2	0	0	0	0	-2	2	0	0	0	0

The IRs of G^+_{16} are correlated with those of the fixed-point covering symmetry groups D_{4h} ($\gamma = 0$) and D_{4d} ($\gamma = \pi/4$) in Figure 4.11 (cf. Appendix 4). It can be seen that use of semi-direct product notation for IRs allows very straightforward construction of the rigid/nonrigid correlation diagram. The character table for the single PI group G_{16} can also be readily obtained from Table 4.5.

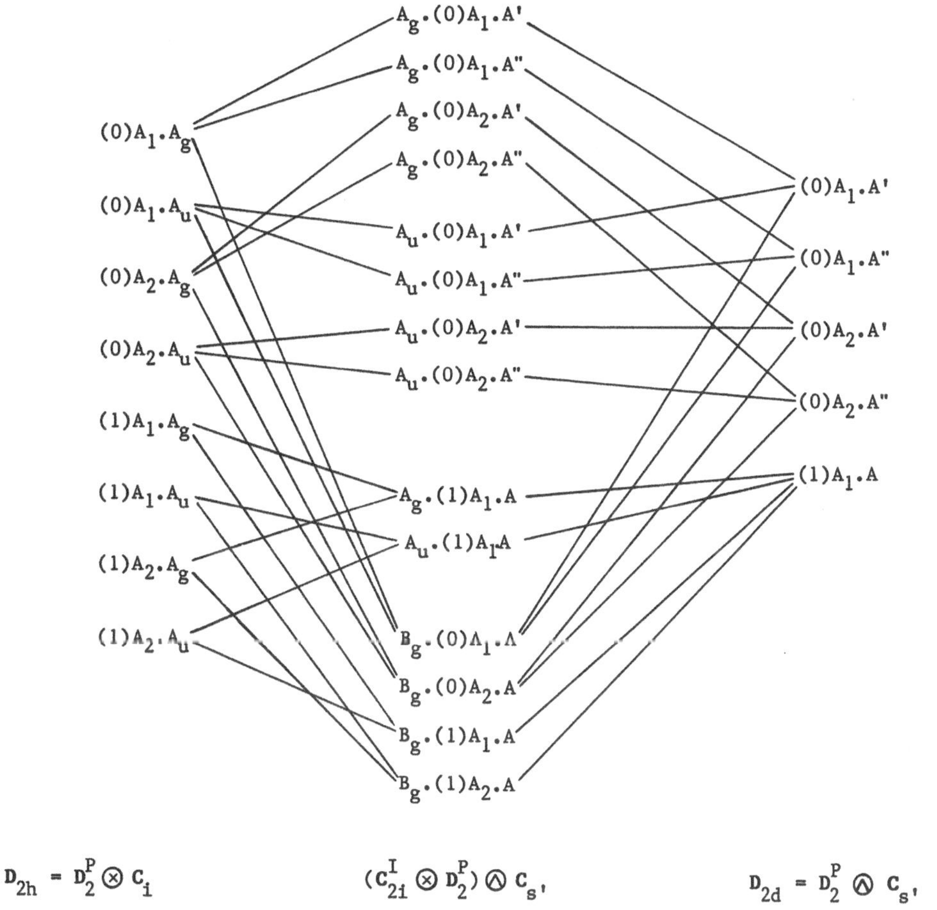

Figure 4.11. Correlation of the IRs of $(C^I_{2i} \otimes D^P_2) \circledA C_{s'}$ with those of D_{2h} and D_{2d}.

Apart from being appropriate for molecules such as $BF_2 - BF_2$ undergoing internal rotation, the group G^+_{16} is also isomorphic with the symmetry group of the inversion/internal-rotor SRMM for hydrazine [158].

b) $n = 4$: $XY_4 - XY_4$

The molecular model for the rotor $XY_4 - XY_4$ is shown in Figure 4.8b; it can be used to describe the dimer $SF_5 - SF_5$, for example. The symmetry group H is of order 128

$$H = (C^{I}_{4i} \otimes D^{P}_{4}) \circledS C_{s'} \quad . \qquad 4.34$$

The character table for this group, which has 28 IRs, is given elsewhere [189]. The character table for the single PI group G_{64} has been determined by Bunker [96]; however, the double group H is required when considering the symmetry properties of the molecular rotational, vibrational and torsional wavefunctions separately (cf. §3.2).

c) $n = 6$: $XY_6 - XY_6$

The molecular model for $n = 6$ is shown in Figure 4.8c, and can be used to describe the essentially unhindered internal rotations in sandwich compounds such as dibenzene chromium.

The symmetry group of the SRMM is a semi-direct product of order 288:

$$H = (C^{I}_{6i} \otimes D^{P}_{6}) \circledS C_{s'} = G^{+}_{144} \quad , \qquad 4.35$$

having 42 IRs; its character table is presented elsewhere [189].

4.4 Some other rotors

In this section we consider the symmetry groups for three rotor NRMs having molecular models described by more than one large-amplitude parameter.

a) $XY_3 - Z - XY_3$

This molecular model is shown in Figure 4.12, and describes internal rotations in both dimethyl ether ($CH_3 - O - CH_3$) [96] and propane ($CH_3 - CH_2 - CH_3$) [159, 160].

There are two SRMM parameters, the torsional angles (γ_1, γ_2) which have ranges $0 \leq \gamma_1, \gamma_2 \leq 2\pi$. The intrinsic group is of order 9 and is isomorphic with the direct product of two cyclic groups of order 3:

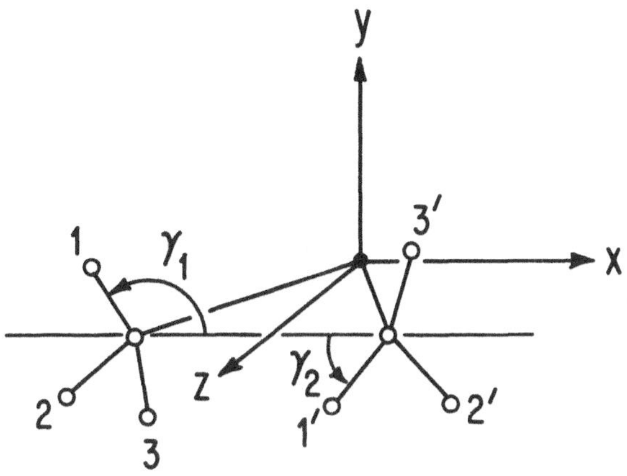

Figure 4.12 The SRMM for XY_3 -Z - XY_3

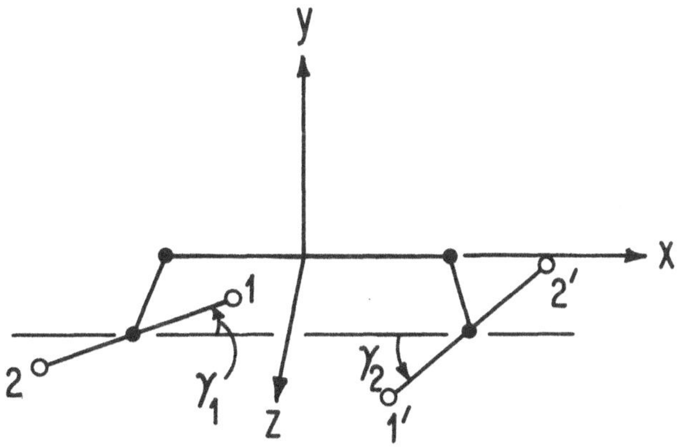

Figure 4.14 The SRMM for XY_2 - Z -Z -XY_2

$$G^I = C_3^I(1) \otimes C_3^I(2) \qquad 4.36$$

The intrinsic group G^I has elements

$$(\mu_1)(\mu_2) \equiv h_1^{\mu_1} \cdot h_2^{\mu_2} \qquad \mu_1, \mu_2 \text{ mod } 3 \qquad 4.37a$$

where the generators h_1, h_2 are

$$h_1 \equiv (\hat{E}, (\gamma_1 + 2\pi/3, \gamma_2)) \qquad 4.37b$$

$$h_2 \equiv (\hat{E}, (\gamma_1, \gamma_2 + 2\pi/3)) , \qquad 4.37c$$

and IRs denoted

$$(m_1)(m_2) \qquad m_1, m_2 \ \underline{\text{mod}}\ 3 \ . \qquad 4.38$$

The factor group $\mathbf{K}$ is isomorphic with C_{2v},

$$\mathbf{K} = C_{2v} = \{E, C_{2v}, \sigma, \sigma'\} \qquad 4.39a$$

$$E \equiv (\hat{E}, (\gamma_1, \gamma_2)) \qquad 4.39b$$

$$C_{2z} \equiv (\hat{C}_{2z}, (\gamma_2, \gamma_1)) \qquad 4.39c$$

$$\sigma \equiv (\hat{\sigma}_{yz}, (-\gamma_2, -\gamma_1)) \qquad 4.39d$$

$$\sigma' \equiv (\hat{\sigma}_{xz}, (-\gamma_1, -\gamma_2)) , \qquad 4.39e$$

so that the symmetry group of the SRMM is the semi-direct product

$$\mathbf{H} = (C_3^I(1) \otimes C_3^I(2)) \circledS C_{2v} \qquad 4.40$$

of order 36 [159].

The class structure of $\mathbf{H}$ is as follows.

Order	
1	$[E\|(0)(0)]$
4	$[E\|(1)(0)]$, $[E\|(0)(1)]$, $[E\|(2)(0)]$, $[E\|(0)(2)]$
2	$[E\|(1)(2)]$, $[E\|(2)(1)]$
2	$[E\|(1)(1)]$, $(E\|(2)(2)]$
3	$[C_{2z}\|(0)(0)]$, $(C_{2z}\|(1)(2)]$, $[C_{2z}\|(2)(1)]$
6	$[C_{2z}\|(1)(0)]$, $[C_{2z}\|(0)(1)]$, $[C_{2z}\|(2)(0)]$ $[C_{2z}\|(0)(2)]$, $[C_{2z}\|(1)(1)]$, $[C_{2z}\|(2)(2)]$
3	$[\sigma\|(0)(0)]$, $[\sigma\|(1)(1)]$, $[\sigma\|(2)(2)]$
6	$[\sigma\|(0)(1)]$, $[\sigma\|(1)(0)]$, $[\sigma\|(0)(2)]$ $[\sigma\|(2)(0)]$, $[\sigma\|(1)(2)]$, $[\sigma\|(2)(1)]$
9	$[\sigma'\|(0)(0)]$, $[\sigma'\|(0)(1)]$, $[\sigma'\|(0)(2)]$ $[\sigma'\|(1)(0)]$, $[\sigma'\|(2)(0)]$, $[\sigma'\|(1)(1)]$ $[\sigma'\|(1)(2)]$, $[\sigma'\|(2)(1)]$, $[\sigma'\|(2)(2)]$

and orbits of $C_3^I(1) \otimes C_3^I(2)$ under C_{2v} are

Orbit	Little Co-group
$\{(0)(0)\}$	C_{2v}
$\{(1)(1), (2)(2)\}$	$C_2(z)$
$\{(2)(1), (2)(1)\}$	$C_s(\sigma)$
$\{(1)(0), (0)(1), (2)(0), (0)(2)\}$	$\{E\}$.

The character table for $(C_3^I \otimes C_3^I) \circledS C_{2v}$ is given in Table 4.6.

Table 4.6. Character table for $(C_3^I \otimes C_3^I) \circledwedge C_{2v}$

	1	4	2	2	3	6	3	6	9
	$[E\|(0)(0)]$	$[E\|(1)(0)]$	$[E\|(1)(1)]$	$[E\|(1)(2)]$	$[C_{2z}\|(0)(0)]$	$[C_{2z}\|(1)(0)]$	$[\sigma_v\|(0)(0)]$	$[\sigma_v\|(0)(1)]$	$[\sigma_v'\|(0)(0)]$
$(0)(0).A_1$	1	1	1	1	1	1	1	1	1
$(0)(0).A_2$	1	1	1	1	1	1	-1	-1	-1
$(0)(0).B_1$	1	1	1	1	-1	-1	1	1	-1
$(0)(0).B_2$	1	1	1	1	-1	-1	-1	-1	1
$(1)(1).A$	2	-1	-1	2	2	-1	0	0	0
$(1)(1).B$	2	-1	-1	2	-2	1	0	0	0
$(2)(1).A_1$	2	-1	2	-1	0	0	2	-1	0
$(2)(1).A_2$	2	-1	2	-1	0	0	-2	1	0
$(1)(0).A$	4	1	-2	-2	0	0	0	0	0

IRs of $(C_3^I \otimes C_3^I) \circledwedge C_{2v}$ are correlated with those of its subgroup C_{2v} in Figure 4.13; a portion of this correlation diagram has previously been given by Flurry [159]. The NRM group derived here is isomorphic with the PI group[96], and is equivalent to the 'isodynamic' group obtained by Flurry [159].

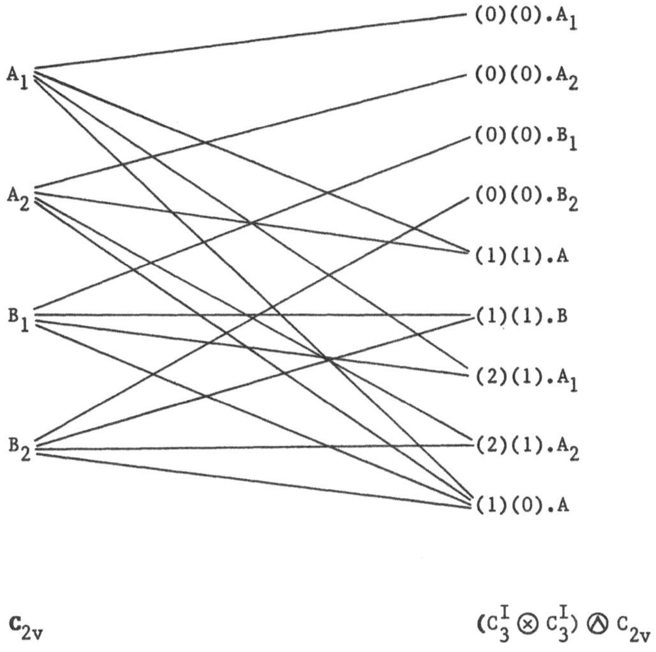

Figure 4.13. Correlation of the IRs of $(C_3^I \otimes C_3^I) \circledwedge C_{2v}$ with those of C_{2v}

b) $XY_2 - Z - Z - XY_2$

The two-parameter model for $XY_2 - Z - Z - XY_2$ is shown in Figure 4.14; it has been used by Wallace [162] as a basis for describing the butadiene - cyclobutane rearrangement.

The two SRMM parameters are the torsional angles (γ_1, γ_2), where $0 \leqslant \gamma_1, \gamma_2 \leqslant 2\pi$. The intrinsic group is of order 4

$$G^{I} = C_2^{I}(1) \otimes C_2^{I}(2) \qquad 4.41$$

with elements

$$(\mu_1)(\mu_2) \equiv h_1^{\mu_1} \cdot h_2^{\mu_2} \qquad \mu_1,\mu_2 \ \underline{\text{mod}}\ 2 \qquad 4.42a$$

$$h_1 \equiv (\hat{E},(\gamma_1 + \pi,\ \gamma_2)) \qquad 4.42b$$

$$h_2 \equiv (\hat{E},(\gamma_1,\ \gamma_2 + \pi)) \qquad 4.42c$$

and IRs denoted

$$(m_1)(m_2) \qquad m_1,m_2\ \underline{\text{mod}}\ 2\ . \qquad 4.43$$

The group **K** is isomorphic with C_{2v}, and with the definitions of axes and torsional angles shown in Figure 4.14 is identical with the C_{2v} group defined in equation 4.39. **H** is then the semi-direct product

$$\mathbf{H} = (C_2^{I} \otimes C_2^{I}) \circledS C_{2v} \qquad 4.44$$

of order 16.

Orbits of $(C_2^{I} \otimes C_2^{I})$ under C_{2v} are

Orbit	Little Co-group
{(0)(0)}	C_{2v}
{(1)(1)}	C_{2v}
{(1)(0), (0)(1)}	$C_s(\sigma)$,

and the character table for $(C_2^{I} \otimes C_2^{I}) \circledS C_{2v}$ is shown in Table 4.7.

Table 4.7. Character table for $(C_2^I \otimes C_2^I) \otimes C_{2v}$

	1	1	2	1	1	2	2	2	2	2
	$[E\vert(0)(0)]$	$[E\vert(1)(1)]$	$[E\vert(1)(0)]$	$[\sigma\vert(0)(0)]$	$[\sigma\vert(1)(1)]$	$[\sigma\vert(1)(0)]$	$[C_{2z}\vert(0)(0)]$	$[C_{2z}\vert(0)(1)]$	$[\sigma'\vert(0)(0)]$	$[\sigma'\vert(0)(1)]$
$(0\vert(0).A_1$	1	1	1	1	1	1	1	1	1	1
$(0\vert(0).A_2$	1	1	1	-1	-1	-1	1	1	-1	-1
$(0\vert(0).B_1$	1	1	1	1	1	1	-1	-1	-1	-1
$(0\vert(0).B_2$	1	1	1	-1	-1	-1	-1	-1	1	1
$(1\vert(1).A_1$	1	1	-1	1	1	-1	1	-1	1	-1
$(1\vert(1).A_2$	1	1	-1	-1	-1	1	1	-1	-1	1
$(1\vert(1).B_1$	1	1	-1	1	1	-1	-1	1	-1	1
$(1\vert(1).B_2$	1	1	-1	-1	-1	1	-1	1	1	-1
$(1\vert(0).A_1$	2	-2	0	2	-2	0	0	0	0	0
$(1\vert(0).A_2$	2	-2	0	-2	2	0	0	0	0	0

In order to model the concerted butadiene-cyclobutadiene rearrangement, Wallace has considered the abelian subgroup of $\mathbf{H}$ of order 8 describing internal motion of the rotors in phase [162]

$$(C_2^I \circledcirc{d}\; C_2^I) \circledwedge C_{2v} \overset{\text{iso}}{\equiv} C_2 \otimes C_{2v} \subset \mathbf{H} \,. \tag{4.45}$$

The group $(C_2^I \otimes C_2^I) \circledwedge C_{2v}$ is also appropriate for treating internal rotations of the aryl groups in diphenylmethane [163].

c) Tetraphenylmethane

The molecular model for the NRM tetraphenylmethane, $C(C_6H_5)_4$, is shown in Figure 4.15. As mentioned in Chapter 3, the symmetry properties of this molecule have recently been discussed by Nourse and Mislow [145].

The SRMM depends upon four parameters, the torsional angles $(\gamma_1, \gamma_2, \gamma_3, \gamma_4)$ defined in Figure 4.15. The intrinsic group $\mathbf{G}^I$ is a direct product of four 2-fold ring rotation groups

$$\mathbf{G}^I = C_2^I(1) \otimes C_2^I(2) \otimes C_2^I(3) \otimes C_2^I(4) \tag{4.46}$$

with elements and IRs denoted

$$(\mu_1, \mu_2, \mu_3, \mu_4) \qquad \mu_1, \mu_2, \mu_3, \mu_4 \ \underline{\text{mod}}\ 2 \tag{4.47}$$

in our standard notation.

Defining the operations (the C_3-axis lies along the direction (1,1,1))

$$E \equiv (\hat{E}, (\gamma_1, \gamma_2, \gamma_3, \gamma_4)) \tag{4.48a}$$

$$C_{2x} \equiv (\hat{C}_{2x}, (\gamma_3 + \pi, \gamma_4 + \pi, \gamma_1 + \pi, \gamma_2 + \pi)) \tag{4.48b}$$

$$C_{2y} \equiv (\hat{C}_{2y}, (\gamma_2 + \pi, \gamma_1 + \pi, \gamma_4 + \pi, \gamma_3 + \pi)) \tag{4.48c}$$

$$C_{2z} \equiv (\hat{C}_{2z}, (\gamma_4, \gamma_3, \gamma_2, \gamma_1)) = C_{2x} \cdot C_{2y} \tag{4.48d}$$

$$C_3^+ \equiv (\hat{C}_3^+, (\gamma_3 - \pi/3, \gamma_1 - \pi/3, \gamma_2 + 2\pi/3, \gamma_4 + 2\pi/3)) \tag{4.48e}$$

$$\sigma_d \equiv (\hat{\sigma}_d, (-\gamma_1, -\gamma_3, -\gamma_2, -\gamma_4)) \tag{4.48f}$$

and groups

$$\mathbf{D}_2 \equiv \{E, C_{2x}, C_{2y}, C_{2z}\} \tag{4.49a}$$

$$\mathbf{C}_3 \equiv \{E, C_3^+, (C_3^+)^2\} \tag{4.49b}$$

$$\mathbf{C}_s \equiv \{E, \sigma_d\} \tag{4.49c}$$

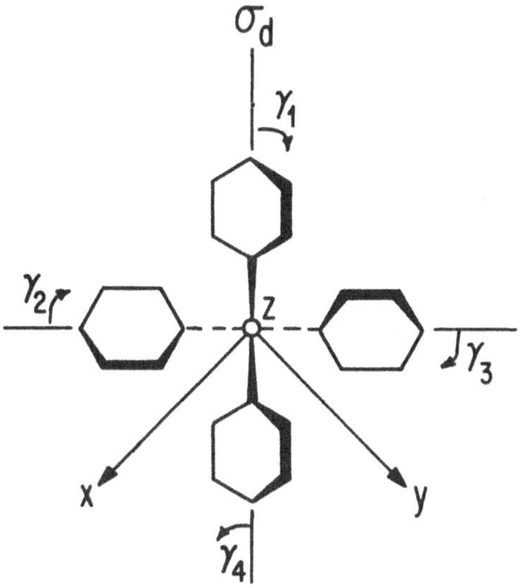

Figure 4.15 The SRMM for tetraphenylmethane

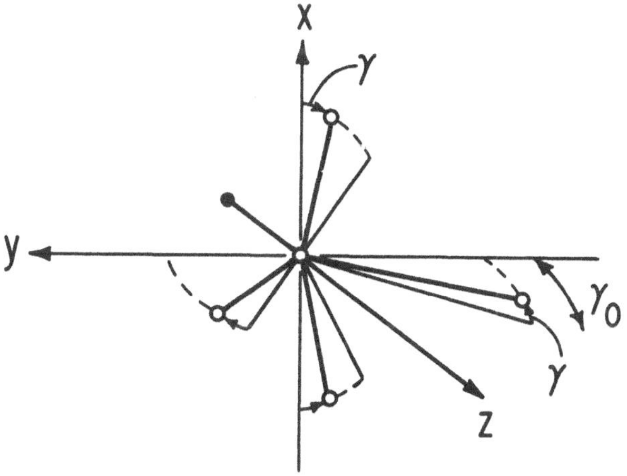

Figure 4.16 The SRMM for pseudorotating ZXY_4

$$C_{3v} \equiv C_3 \circledwedge C_s \qquad 4.49d$$

$$K \overset{iso}{\equiv} T_d = D_2 \circledwedge C_{3v} \qquad 4.49e$$

the symmetry group of the tetraphenylmethane model can be written as the semi-direct product

$$H = (C_2^I \otimes C_2^I \otimes C_2^I \otimes C_2^I) \circledwedge T_d \equiv \text{"}G_{384}\text{"} \qquad 4.50$$

of order 384. As noted in Chapter 3 the tetraphenylmethane molecule has no proper covering symmetry for arbitrary values of the parameters $(\gamma_1, \gamma_2, \gamma_3, \gamma_4)$, so that the point group G^P is the trivial group C_1. However, the effective point symmetry K induced by the nonrigidity (ring rotation) is the tetrahedral group T_d.

Orbits of the invariant subgroup G^I under T_d are

Orbit	Little Co-group
{(0,0,0,0)}	$\mathcal{K}(0,0,0,0) = T_d$
{(0,0,0,1), (0,0,1,0), (0,1,0,0), (1,0,0,0)}	$\mathcal{K}(0,0,0,1) = C_{3v}$
{(1,0,0,1), (0,1,0,1), (0,0,1,1), (1,0,1,0), (0,1,1,0), (1,1,0,0)}	$\mathcal{K}(1,0,0,1) = C_2(z) \circledwedge C_s$
{(1,1,1,0), (1,1,0,1), (1,0,1,1), (0,1,1,1)}	$\mathcal{K}(1,1,1,0) = C_{3v}$
{(1,1,1,1)}	$\mathcal{K}(1,1,1,1) = T_d$

where, for example, $\mathcal{K}(0,0,0,1) = C_{3v}$ is the little co-group associated with the particular IR (0,0,0,1) of the invariant subgroup.

The character table for G_{384} is shown in Table 4.8 (the complete class structure is not given here). As pointed out by Nourse and Mislow [145], the character table for G_{384}, which is isomorphic with a transitive subgroup of the symmetric group of degree 8, has previously been given by Littlewood ([161], p. 278). As well as serving as an independent check, our calculation produces the table in a form that best displays its structure. (It is also of interest to note that the character table for the symmetry group of the NRM $B(CH_3)_3$ [1], which is the semi-direct product $(C_3^I \otimes C_3^I \otimes C_3^I) \circledwedge D_{3h}$ of order 324 [125] isomorphic with a

Table 4.8. Character table for $(C_2^I \otimes C_2^I \otimes C_2^I \otimes C_2^I) \circledS T_d$, symmetry group of the NRM tetraphenylmethane.

	1	4	6	4	1	12	24	12	32	32	32	32	12	24	12	12	24	12	48	48
	E	E	E	E	E	C_{2z}	C_{2z}	C_{2z}	C_3	C_3	C_3	C_3	σ_d	σ_d	σ_d	σ_d	σ_d	σ_d	S_4	S_4
	(0000)	(0001)	(0011)	(0111)	(1111)	(0000)	(0001)	(0011)	(0000)	(0001)	(0010)	(0011)	(0000)	(1010)	(1111)	(1101)	(0001)	(0010)	(0000)	(0001)
(0000].A_1	1	1	1	1	1	1	1	1	1	1	1	1	1	1	1	1	1	1	1	1
(0000].A_2	1	1	1	1	1	1	1	1	1	1	1	1	-1	-1	-1	-1	-1	-1	-1	-1
(0000].E	2	2	2	2	2	2	2	2	-1	-1	-1	-1	0	0	0	0	0	0	0	0
(0000].T_2	3	3	3	3	3	-1	-1	-1	0	0	0	0	1	1	1	1	1	1	-1	-1
(0000].T_1	3	3	3	3	3	-1	-1	-1	0	0	0	0	-1	-1	-1	-1	-1	-1	1	1
(1111].A_1	1	-1	1	-1	1	1	-1	1	1	-1	-1	1	1	1	1	-1	-1	-1	1	-1
(1111].A_2	1	-1	1	-1	1	1	-1	1	1	-1	-1	1	-1	-1	-1	1	1	1	-1	1
(1111].E	2	-2	2	-2	2	2	-2	2	-1	1	1	-1	0	0	0	0	0	0	0	0
(1111].T_2	3	-3	3	-3	3	-1	1	-1	0	0	0	0	1	1	1	-1	-1	-1	-1	1
(1111].T_1	3	-3	3	-3	3	-1	1	-1	0	0	0	0	-1	-1	-1	1	1	1	1	-1
(0001].A_1	4	2	0	-2	-4	0	0	0	1	-1	1	-1	2	0	-2	-2	0	2	0	0
(0001].A_2	4	2	0	-2	-4	0	0	0	1	-1	1	-1	-2	0	2	2	0	-2	0	0
(0001].E	8	4	0	-4	-8	0	0	0	-1	1	-1	1	0	0	0	0	0	0	0	0
(1110].A_1	4	-2	0	2	-4	0	0	0	1	1	-1	-1	2	0	-2	2	0	-2	0	0
(1110].A_2	4	-2	0	2	-4	0	0	0	1	1	-1	-1	-2	0	2	-2	0	2	0	0
(1110].E	8	-4	0	4	-8	0	0	0	-1	-1	1	1	0	0	0	0	0	0	0	0
(1001].A_1	6	0	-2	0	6	2	0	-2	0	0	0	0	2	-2	2	0	0	0	0	0
(1001].A_2	6	0	-2	0	6	2	0	-2	0	0	0	0	-2	2	-2	0	0	0	0	0
(1001].B_1	6	0	-2	0	6	-2	0	2	0	0	0	0	0	0	0	2	-2	2	0	0
(1001].B_2	6	0	-2	0	6	-2	0	2	0	0	0	0	0	0	0	-2	2	-2	0	0

transitive subgroup of the symmetric group of degree 9, is given in Littlewood's book as well: [161], p.282).

4.5 Pseudorotation of ZXY_4.

The last NRM we consider here is the pseudorotating ZXY_4 molecule, whose 1-parameter SRMM is shown in Figure 4.16. The presence of the pivot ligand Z, which is assumed to remain in an equatorial position, avoids the full complexity of the XY_5 problem (cf. §1.5). Pseudorotation is represented by the transformation

$$\tau:\ \gamma \rightarrow \gamma_0 - \gamma \qquad 4.69$$

(the dynamics of this system have been studied by Russeger and Brickmann [55]).

The point group of the SRMM is C_{2v}

$$\mathbf{G}^p \stackrel{iso}{=} C_{2v} \equiv \{E, C_{2z}, \sigma_{yz'}, \sigma_{xz}\}\ . \qquad 4.70$$

Defining the group of order 2

$$\mathbf{K} \stackrel{iso}{=} C_{s'} = \{(\hat{E},\gamma),\ (\hat{\sigma}',\gamma_0 - \gamma)\} \qquad 4.71$$

(where the reflection plane σ' contains the z-axis and bisects the xy-plane), the symmetry group of the SRMM is the semi-direct product

$$\mathbf{H} = C_{2v} \circledS C_{s'} \stackrel{iso}{=} C_{4v} \qquad 4.72$$

of order 8, and has the character table shown in Table 4.9, This table agrees with that obtained by Dalton [68] for the Q-group.

The symmetry labels for the states of a rigid C_{2v} molecule are correlated with those of the pseudorotating NRM in Figure 4.17.

Table 4.9. Character table of $C_{2v} \circledcirc C_{s'}$

$C_{2v} \circledcirc C_{s'}$	[E|E]	[E|C_{2z}]	[E|σ_{yz}] [E|σ_{xz}]	[σ'|C_{2z}] [σ'|E]	[σ'|σ_{yz}] [σ'|σ_{xz}]
(0)$A_1 \cdot A'$	1	1	1	1	1
(0)$A_1 \cdot A''$	1	1	1	-1	-1
(0)$A_2 \cdot A'$	1	1	-1	1	-1
(0)$A_2 \cdot A''$	1	1	-1	-1	1
(1)$A_1 \cdot A$	2	-2	0	0	0

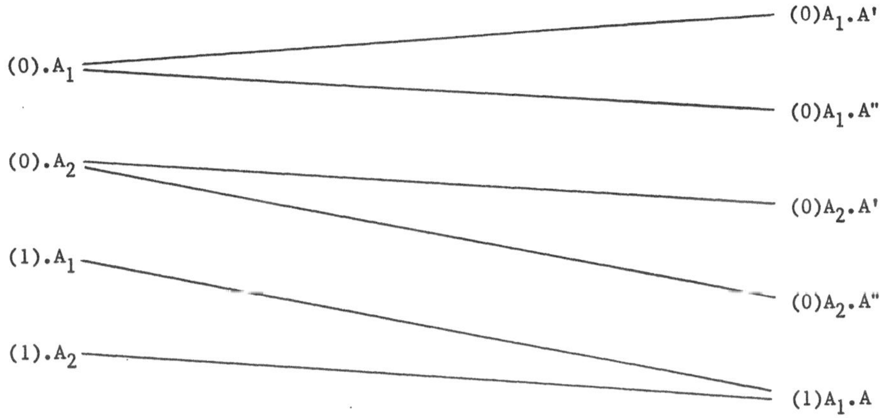

Figure 4.17. Correlation of the IRs of C^P_{2v} with those of $C_{2v} \circledcirc C_{s'}$

4.6 Conclusions and Possibilities for Future Developments

We have presented in this chapter several applications of our formalism to the study of the symmetry properties of nonrigid molecules. We mention here some possibilities for future work:

i) The semi-direct product decomposition of the symmetry group of the semi-rigid molecular model is clearly significant for the classification of molecular wavefunctions and operators in the nonrigid molecule group. Thus, the molecule-fixed components of multipole tensors are necessarily invariant under operations of the intrinsic group, and so need only be classified in the group $\mathbf{G}^P \circledwedge \mathbf{K}$, which is the effective point symmetry group of the molecule. It is therefore possible to utilize our knowledge of the transformation properties of multipole tensors in point symmetry groups [83, 119] when considering their classification in NRM groups (cf. [124]). In fact, one might hope to formulate a theory of selection rules in NRMs on the basis of semi-direct product structure; such a scheme would be very close in spirit to recent developments in the use of double-coset algebras [179, 180].

ii) The treatment of the symmetry aspects of nuclear vibrations in NRMs has generally proceeded in a fairly <u>ad hoc</u> fashion. Some steps in the direction of a more systematic treatment have been taken by Natanson [126, 154, 183], who has introduced the useful notion of 'nontrivial' operations of the NRM group. In our terminology nontrivial operations are either elements of the group $\mathbf{K}$ that do not become fixed-point covering operations or elements of the intrinsic group $\mathbf{G}^I$, having the property that there is a one-dimensional irreducible representation of the nonrigid molecule group for which the corresponding character is −1. An important aspect of the general theory of nuclear vibrations involves the correlation of nonrigid molecule group irreducible representations with those of covering symmetry groups corresponding to various fixed points in parameter space. The correlation of nonrigid molecule irreducible representations with those of the point group $\mathbf{G}^P$ itself is also important. As we have sought to emphasize throughout this chapter, the semi-direct product formulation of nonrigid molecule symmetry groups is ideally suited to the construction of the appropriate rigid/nonrigid correlation diagrams. There is a need for more work along these lines. (See, for example, the very recent work by Renkes [191] on the determination of symmetry groups for systems with motions described by several large-amplitude coordinates. While emphasizing the process of subduction rather than induction, this analysis is close in spirit to that we have presented here, since it deals directly with the transformations of internal parameters corresponding to various symmetry operations. The symmetries may be either exact or approximate (neglect of coupling terms)).

iii) An account of the electron spin 'double' groups of NRM symmetry groups has recently been given by Bunker [164]. The effects of electron spin can be treated very simply in terms of our formalism. Thus, 'double-valued' spin functions transform as projective irreducible representations of the effective point symmetry group $G^P \otimes K$. Various possible couplings of the electron spin to different parts of the molecule correspond to automorphisms of H induced by changes in the definition of the coordinate frame of the molecule model.

Appendix 1

Rotations of Axes and the Group O(3)

In this appendix we present a brief summary of ideas concerning rotations of axes and coordinate transformations necessary for applications to molecular symmetry. Conventions and notation used throughout our work are introduced.

Consider the vector space $\mathbf{E}$ spanned by an orthonormal basis $\hat{\underset{\sim}{e}}_i$, $i = x,y,z$. The rotation-inversion group O(3) is the set of all mappings ρ of the space $\mathbf{E}$ onto itself such that the scalar product of any two vectors is invariant:

$$O(3) = \{\rho:\mathbf{E} \to \mathbf{E} | \underset{\sim}{a}\cdot\underset{\sim}{b} = \underset{\sim}{a}'\cdot\underset{\sim}{b}'\} \qquad \text{A1.1}$$

where $\rho:\underset{\sim}{a} \to \underset{\sim}{a}'$, $\underset{\sim}{b} \to \underset{\sim}{b}'$ and $\underset{\sim}{a}\cdot\underset{\sim}{b} = a_i b_i$ (note implied summation over component indices i). Operations of O(3) can be defined by their action on the basis vectors $\hat{\underset{\sim}{e}}_i$, viz.,

$$\rho:\hat{\underset{\sim}{e}}_i \to \hat{\underset{\sim}{e}}'_i \equiv \hat{\underset{\sim}{e}}_j R(\rho)_{ji} \qquad \text{A1.2}$$

where the 3 by 3 matrices $R(\rho)$ form the defining or Cartesian representation. The matrices $R(\rho)$ are orthogonal

$$R(\rho)\tilde{R}(\rho) = 1 \qquad \text{A1.3}$$

($\tilde{R}$ is the transpose of R).

Rotations and rotation-inversions of basis vectors defined above induce transformations of vector components; thus, for all $\underset{\sim}{a} = \hat{\underset{\sim}{e}}_i a_i$ in $\mathbf{E}$

$$\rho:\underset{\sim}{a} \to \underset{\sim}{a} = \hat{\underset{\sim}{e}}_i a'_i \qquad \text{A1.4a}$$

with

$$a'_i = R(\rho)_{ij} a_j \quad . \qquad \text{A1.4b}$$

This corresponds to the 'active' convention for rotations, in which the vector $\underset{\sim}{a}$ is 'rotated with the basis', and new components a'_i are calculated with respect to the original basis (cf. Figure A1.1).

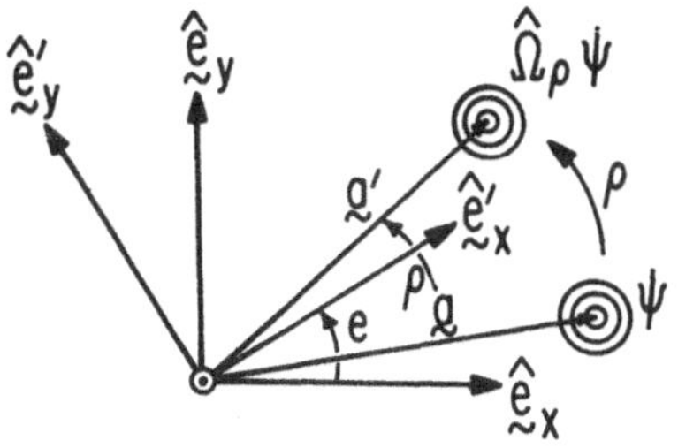

Figure A1.1 The effect of a rotation of axes upon the vector $\underset{\sim}{a}$ and the scalar ψ .

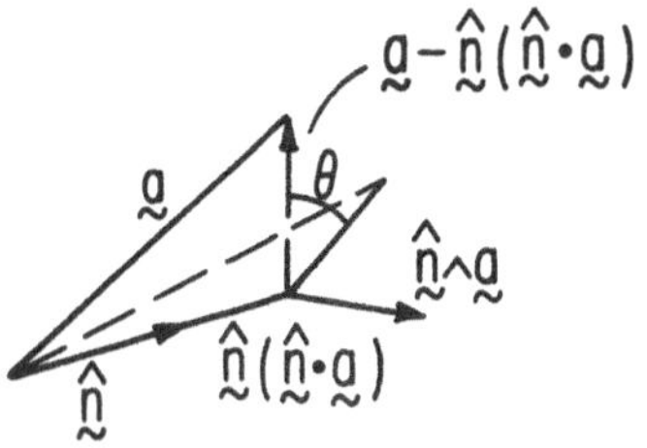

Figure A1.2 Rotation of the vector $\underset{\sim}{a}$ through θ about the axis $\hat{\underset{\sim}{n}}$.

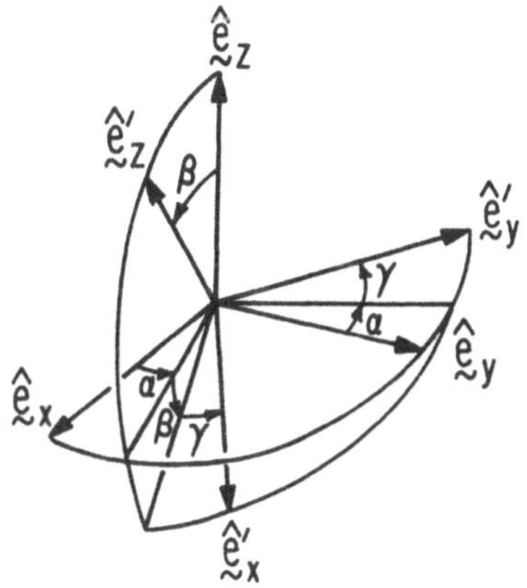

Figure A1.3 Transformation of the coordinate frame $\{\hat{\underset{\sim}{e}}_i\}$ to $\{\hat{\underset{\sim}{e}}'_i\}$ specified by the Euler angles (α, β, γ).

The vector product of any 2 vectors $\underset{\sim}{a}$ and $\underset{\sim}{b}$ is

$$\underset{\sim}{a} \wedge \underset{\sim}{b} = \hat{\underset{\sim}{e}}_i \, \varepsilon_{ijk} a_j b_k \, , \qquad \text{A1.5}$$

i.e., the 'right-hand screw rule', and is invariant under the inversion $\underset{\sim}{a} \to -\underset{\sim}{a}$, $\underset{\sim}{b} \to -\underset{\sim}{b}$.

The orthogonality relations A1.3 constitute 6 conditions on the matrix elements R_{ij}, so that there are 3 independent degrees of freedom necessary for specification of a matrix $R(\rho)$. At some point a particular parametrization must be adopted. A scheme well suited for use in point group theory defines (for proper rotations) the orientation of a rotation axis $\hat{\underset{\sim}{n}}$ (unit vector in E) together with an angle of rotation θ $(0 \leq \theta \leq \pi)$ through which rotation occurs about $\hat{\underset{\sim}{n}}$ in a positive sense. Restriction of the range of θ to $0 \leq \theta \leq \pi$ allows a unique association of proper rotations with pairs $(\theta,\hat{\underset{\sim}{n}})$, except that $\rho(\pi,\hat{\underset{\sim}{n}}) \equiv \rho(\pi,-\hat{\underset{\sim}{n}})$.

The action of a rotation $\rho(\theta,\hat{\underset{\sim}{n}})$ on the basis vectors is (Figure A1.2)

$$\rho(\theta,\hat{\underset{\sim}{n}}):\hat{\underset{\sim}{e}}_j \to \hat{\underset{\sim}{n}}(\hat{\underset{\sim}{n}} \cdot \hat{\underset{\sim}{e}}_j) + \cos\theta(\hat{\underset{\sim}{e}}_j - \hat{\underset{\sim}{n}}(\hat{\underset{\sim}{n}} \cdot \hat{\underset{\sim}{e}}_j)) + \sin\theta(\hat{\underset{\sim}{n}} \wedge \hat{\underset{\sim}{e}}_j)$$

$$= \hat{\underset{\sim}{e}}_i [n_i(1 - \cos\theta)n_j + \cos\theta\, \delta_{ij} - \sin\theta\, \varepsilon_{ijk} n_k] \qquad \text{A1.6}$$

with $\hat{\underset{\sim}{n}} = n_i \hat{\underset{\sim}{e}}_i$. The matrix $R(\rho)$ is therefore

$$R(\theta,\hat{\underset{\sim}{n}})_{ij} = n_i(1 - \cos\theta)n_j + \cos\theta\, \delta_{ij} - \sin\theta\, \varepsilon_{ijk} n_k \, . \qquad \text{A1.7}$$

Note that improper rotations can always be written as the product of the inversion operation $\hat{\underset{\sim}{e}}_i \to -\hat{\underset{\sim}{e}}_i$ and a proper rotation ρ .

It is usual to define proper rotations of axes in terms of 3 Euler angles defined as in Figure A1.3. The rotation of axes denoted $\rho(\alpha,\beta,\gamma)$ is the product

$$\rho(\alpha,\beta,\gamma) = \rho(\alpha,\hat{\underset{\sim}{e}}_z)\rho(\beta,\hat{\underset{\sim}{e}}_y)\rho(\gamma,\hat{\underset{\sim}{e}}_z), \qquad \text{A1.8}$$

where the order of rotations should be noted $(\rho(\gamma,\hat{\underset{\sim}{e}}_z)$ acts first), as should the fact that axes of rotation are defined with respect to the original basis. Parameter ranges are $0 \leq \alpha \leq 2\pi$, $0 \leq \beta \leq \pi$, $0 \leq \gamma \leq \pi$, and implicit use is made of the identity $\rho(\theta,-\hat{\underset{\sim}{n}}) = \rho(2\pi - \theta,\hat{\underset{\sim}{n}})$. From the general result A1.7 we have

$$R(\alpha,\beta,\gamma) = R(\alpha,\hat{\underset{\sim}{e}}_z)R(\beta,\hat{\underset{\sim}{e}}_y)R(\gamma,\hat{e}_z) \qquad \text{A1.9a}$$

$$= \begin{bmatrix} \cos\alpha\cos\beta\cos\gamma - \sin\alpha\sin\gamma & -\cos\alpha\cos\beta\sin\gamma - \sin\alpha\cos\gamma & \cos\alpha\sin\beta \\ \sin\alpha\cos\beta\cos\gamma + \cos\alpha\sin\gamma & -\sin\alpha\cos\beta\sin\gamma + \cos\alpha\cos\gamma & \sin\alpha\sin\beta \\ -\sin\beta\cos\gamma & \sin\beta\sin\gamma & \cos\beta \end{bmatrix} \qquad \text{A1.9b}$$

which is the familiar form of the direction-cosine matrix in the Euler angle parametrization [165, 166].

Having found coordinate transformations associated with elements of O(3), we can obtain the induced action of group operations upon functions of position. Thus, consider a scalar field $\psi(x_i)$. The mapping

$$\rho:\psi \rightarrow \Omega_\rho\psi \quad , \qquad \text{A1.10}$$

where the new function $\hat{\Omega}_\rho\psi$ satisfies the relation [115]

$$[\hat{\Omega}_\rho\psi](R(\rho)_{ij}x_j) = \psi(x_i) \quad , \qquad \text{A1.11a}$$

i.e.

$$[\Omega_\rho\psi](x_i) = \psi(\tilde{R}(\rho)_{ij}x_j) \quad , \qquad \text{A1.11b}$$

defines the effect of a rotation upon a function. An important result is that the mapping corresponding to the operation $\rho_{12} = \rho_1 \cdot \rho_2$ is the product

$$\hat{\Omega}_{\rho_{12}} = \hat{\Omega}_{\rho_1}\hat{\Omega}_{\rho_2} \quad . \qquad \text{A1.12}$$

To obtain this isomorphic operator representation, we must induce the action of group operations upon functions as in equation A1.11, rather than the alternative expression

$$[\hat{\Omega}_\rho\psi](x_i) = \psi(R(\rho)_{ij}x_j) \ ,$$

which leads to an anti-isomorphism (cf. [90]).

Consider a set of n_Γ linearly independent functions $\{\psi_\gamma^\Gamma(x_i);\ \gamma = 1,..n_\Gamma\}$. The action of an element of O(3) upon a member of the set is

$$\rho:\psi^{\Gamma}_{\gamma}(x_i) \to \psi^{\Gamma}_{\gamma}(\tilde{R}(\rho)_{ij}x_j) = \sum_{\gamma'} \psi^{\Gamma}_{\gamma'}(x_i)\mathcal{D}^{\Gamma}(R(\rho))_{\gamma'\gamma} \qquad \text{A1.13}$$

which defines the n_{Γ}-dimensional representation matrix $\mathcal{D}^{\Gamma}(R(\rho))$. The matrices $\mathcal{D}^{\Gamma}$ have the representation property

$$\mathcal{D}^{\Gamma}(R(\rho_{12})) = \mathcal{D}^{\Gamma}(R(\rho_1))\mathcal{D}^{\Gamma}(R(\rho_2)) \qquad \text{A1.14}$$

For given j the 2j+1 angular momentum eigenfunctions $\{Y^j_m(\theta,\phi);\ m = j,j-1,\ldots,-j\}$ span a (2j + 1)-dimensional irreducible representation (IR) of O(3):

$$\rho:Y^j_m \to \sum_{m'} Y^j_{m'}\mathcal{D}^j(R(\rho))_{m'm} \qquad \text{A1.15}$$

where $\mathcal{D}^j$ is the Wigner rotation matrix (our conventions imply that these rotation matrices are identical with those of Brink and Satchler [165]).

The various transformations introduced above are illustrated in Figure A1.1. The operation $\rho \in O(3)$ rotates the basis vector $\hat{e}_i$ to $\hat{e}'_i$. Similarly, the vector $\underset{\sim}{a}$ is rotated actively into $\underset{\sim}{a}'$. The scalar field ψ is represented by contours about the point $\underset{\sim}{a}$. The function mapping $\psi \to \hat{\Omega}_{\rho}\psi$ can then be visualized as follows: imagine the scalar field rotated with the basis to give a new set of contours centred on $\underset{\sim}{a}'$. The function $\hat{\Omega}_{\rho}\psi$ is the new field expressed in terms of vector components referred to the old basis. The condition $[\hat{\Omega}_{\rho}\psi](a'_i) = \psi(a_i)$ expresses the fact that the scalar field rotates with the basis.

Appendix 2

Angular Momentum and Rotational Wavefunctions.

Several topics related to the subject of angular momentum are discussed in this appendix. In particular, it is pointed out that considerable insight into the transformation properties of rotational wavefunctions can be gained by writing them in terms of the elements of the direction-cosine matrix. These results are utilized in Chapter 2, where we consider the symmetry properties of rigid molecules. The ideas presented here are based upon some remarks of Louck [7] and Louck and Galbraith [17], together with Judd's double-tensor formalism [11]. This approach has also recently been developed by Hilico et al. [14] and by Pascaud [15].

A2.1 Angular Momentum Operators.

Louck has shown by direct transformation of the molecular kinetic energy operator into Born-Oppenheimer coordinates that the laboratory-fixed and (minus) the molecule-fixed components of the total orbital angular momentum of a molecule about its centre of mass are respectively [7]

$$\hat{J}_i = -i\varepsilon_{ijk}C_{js}\partial/\partial C_{ks} \qquad \text{A2.1a}$$

and

$$\hat{K}_i = -i\varepsilon_{ijk}\tilde{C}_{js}\partial/\partial\tilde{C}_{ks} \quad , \qquad \text{A2.1b}$$

where the direction-cosine matrix C describes the orientation of the molecule-fixed Eckart frame with respect to the lab frame and is determined by the instantaneous nuclear positions. Expressions A2.1a,b, which are differential operators on the group manifold of a rotation group SO(3), i.e., the matrix C, are independent of the parametrization of the direction-cosine matrix. It is usual to write $\hat{J}_i$ and $\hat{K}_i$ in terms of the Euler angles (e.g. [11], equations 1.22, 1.23); however, the general expressions A2.1 are especially useful for formal manipulations, and greatly facilitate appreciation of the structure of many results. Moreover, the operators $\hat{J}_i$ and $\hat{K}_i$ are invariant under the inversion transformation

$$\mathcal{J}: C \rightarrow -C \qquad \text{A2.2}$$

so that the relative sense of the lab- and molecule-fixed frames is of no consequence. We can therefore deal consistently with improper rotations of molecule-fixed axes.

Observing that the direction-cosine matrix $C_{ii'} \equiv \hat{\underset{\sim}{\ell}}_i \cdot \hat{\underset{\sim}{f}}_{i'}$ has both 'external' (i) and 'internal' (i') indices, it can be seen that the expressions A2.1a for the lab-fixed components of angular momentum involve a summation over, and are therefore independent of, the internal index s, while the molecule-fixed components involve a summation over the external index. We also see that the transformation $C \rightarrow \tilde{C}$ interconverts the $\hat{J}_i$ and $\hat{K}_i$, indicating a formal algebraic symmetry between internal and external indices.

Components of the vectors $\hat{\underset{\sim}{J}}$ and $\hat{\underset{\sim}{K}}$ satisfy normal angular momentum commutation relations [168, 169] and mutually commute

$$[\hat{J}_i, \hat{J}_j] = i\, \varepsilon_{ijk}\, \hat{J}_k \qquad \text{A2.3a}$$

$$[\hat{K}_i, \hat{K}_j] = i\, \varepsilon_{ijk}\, \hat{K}_k \qquad \text{A2.3b}$$

$$[\hat{J}_i, \hat{K}_{i'}] = 0 \quad , \qquad \text{A2.3c}$$

i.e., we have an abstract SO(3) × SO(3) structure. It is easily shown that

$$\tilde{C}_{ii'} \hat{J}_{i'} = \hat{J}_{i'} C_{i'i} = -(\det C)\hat{K}_i \qquad \text{A2.4}$$

so that

$$\hat{J}_i \hat{J}_i = \hat{K}_i \hat{K}_i \quad . \qquad \text{A2.5}$$

Therefore, although the SO(3) × SO(3) algebra A2.3 involves two commuting angular momenta, the magnitudes of the angular momenta are constrained to be equal through A2.5 (as is obvious since the $\hat{K}_i$ are simply components of the projection of the orbital angular momentum onto the molecule-fixed frame).

Invoking only the commutation relations A2.3 [167] we define eigenfunctions of the angular momenta $\hat{\underset{\sim}{J}}$ and $\hat{\underset{\sim}{K}}$, denoted $|j;m,k\rangle$, which satisfy

$$\hat{J}_z |j;m,k\rangle = m|j;m,k\rangle \qquad \text{A2.6a}$$

$$\hat{K}_z |j;m,k\rangle = k|j;m,k\rangle \qquad \text{A2.6b}$$

$$\hat{J}_\pm |j;m,k\rangle = [j(j+1) - m(m \pm 1)]^{1/2} \, |j; m \pm 1, k\rangle \qquad \text{A2.6c}$$

$$\hat{K}_\pm |j;m,k\rangle = [j(j+1) - k(k \pm 1)]^{1/2} \, |j;m, k \pm 1\rangle \qquad \text{A1.6d}$$

$$\hat{J}_i \hat{J}_i |j;m,k\rangle = \hat{K}_i \hat{K}_i |j;m,k\rangle = j(j+1)|j;m,k\rangle \qquad \text{A2.6e}$$

where $$\hat{J}_\pm = \hat{J}_x \pm i \hat{J}_y \ , \ \hat{K}_\pm = \hat{K}_x \pm \hat{K}_y \ . \qquad \text{A2.7}$$

For given j, the $(2j+1)^2$ functions $\{|j;m,k\rangle; m = j,j-1,\ldots,-j; k = j,j-1,\ldots,-j\}$ form the components of a rotational double-tensor as defined by Judd ([11], equation 1.27). In §A2.3, we consider construction of the $|j;m,k\rangle$ as polynomials in the elements of the matrix C. For the moment, however, we note that the function $|j;m,k\rangle$ span a particular type of IR of the direct product group $SO(3) \times SO(3)$ having equal values of the angular momentum quantum number j and dimension $(2j+1)^2$. Louck and Galbraith have introduced the 'symmetry group of the spherical rotor' $SO^\ell(3) * SO^f(3)$ [17], which has just these IRs

$$SO^\ell(3) \times SO^f(3) \overset{iso}{=} SO^\ell(3) * SO^f(3) \qquad \text{A2.8}$$

(superscripts refer to lab- and frame-fixed operations: see below). In the next section we describe this group and thereby elucidate the group-theoretical significance of the operations $\hat{J}_i$ and $\hat{K}_i$.

A2.2 The Symmetry Group of the Spherical Rotor.

The symmetry group of the spherical rotor $SO^\ell(3) * SO^f(3)$ is the set of all transformations (ρ^ℓ,ρ^f) of the orthogonal matrix C ([17], equation 5.73)

$$(\rho^\ell,\rho^f): C \to R(\rho^\ell)C\tilde{R}(\rho^f) \qquad \text{A2.9}$$

(cf. Appendix 1), with the direct product multiplication rule

$$(\rho_1^\ell,\rho_1^f)(\rho_2^\ell,\rho_2^f) = (\rho_1^\ell\rho_2^\ell,\rho_1^f\rho_2^f). \qquad \text{A2.10}$$

The crucial connection with the results of the previous section is that the generators of the group $SO^\ell(3) * SO^f(3)$ are the angular momentum operators $\hat{J}_i$, $\hat{K}_i$. From A2.9, it is clear that the generators of the subgroups $SO^\ell(3)$ and $SO^f(3)$ ($\hat{J}_i$ and $\hat{K}_i$, respectively) must commute (cf. Equation A2.3c). As pointed out by Louck and Galbraith, this is a global property of the transformation rule A2.9, in that right- and left-multiplication of a given matrix by arbitrary matrices are commuting operations.

It is most important to realize that the group structure $SO^\ell(3) * SO^f(3)$ establishes a clear and rigorous distinction between 'external' (or lab-fixed) and 'internal' (or molecule-fixed) rotations (transformations of the matrix C). Thus, consider the transformation

$$(\rho^\ell,\rho_0^f):C \rightarrow R(\rho^\ell)C \qquad \text{A2.11}$$

where $\rho^\ell \equiv \rho^\ell(\theta^\ell, \hat{\underset{\sim}{n}}^\ell)$ and ρ_0^f is the identity in $SO^f(3)$. This corresponds to the operation: rotate every object in the lab frame, including each particle in the molecule, through an angle θ^ℓ about the axis $\underset{\sim}{n}^\ell$, which is defined as a unit vector fixed with respect to the lab frame $\{\hat{\underset{\sim}{\ell}}_i\}$

$$\hat{\underset{\sim}{n}}^\ell = \hat{\underset{\sim}{\ell}}_i n_i^\ell \quad . \qquad \text{A2.12}$$

This is an <u>external</u> rotation (cf. [96]), and has no induced action upon internal molecular coordinates (as follows from the generalized Malhiot-Ferigle conditions 1.28). The transformation properties of the molecular wavefunction under $SO^\ell(3)$ are therefore entirely determined by the values of the quantum numbers j,m in the rotational wavefunction $|j;k,m\rangle$.

On the other hand, consider the transformation

$$(\rho_0^\ell,\rho^f):C \rightarrow C\tilde{R}(\rho^f) \qquad \text{A2.13}$$

where $\rho^f \equiv \rho^f(\theta^f,\hat{\underset{\sim}{n}}^f)$ and ρ_0^ℓ is the identity in $SO^\ell(3)$. This corresponds to the operation: sitting in the molecule-fixed Eckart frame, rotate every object regarded as <u>external to the molecule</u>, including the lab frame itself, through the angle θ^f

about the axis $\hat{\underset{\sim}{n}}^f$, defined as a unit vector fixed in the Eckart frame $\{\hat{\underset{\sim}{f}}_i\}$

$$\hat{\underset{\sim}{n}}^f = \hat{\underset{\sim}{f}}_i n_i^f \quad . \tag{A2.14}$$

This is an internal rotation, and involves 'rotating the rest of the universe' with respect to the Eckart frame ([83], IIA). Such operations are relevant when we consider the transformation properties of the rotational wavefunctions of molecules possessing covering symmetry (Chapter 2), since the induced action upon rotational coordinates of a permutation of nuclei corresponding to a point group operation g is the frame-fixed rotation $g^f \in SO^f(3)$ (or $O^f(3)$; we consider the extension to $O^f(3) * O^\ell(3)$ below), as should be clear from Chapter 2, equations 2.31a, a'.

It is of course possible to establish a 1:1 correspondence between external and internal rotations. Consider once again the transformation

$$(\rho^\ell, \rho_0^f): C \to R(\rho^\ell)C \quad . \tag{A2.11}$$

For each initial C, there is a corresponding internal transformation $\sigma_\rho^f(C) \in SO^f(3)$ such that

$$C\tilde{R}(\sigma_\rho^f(C)) \equiv R(\rho^\ell)C \ , \tag{A2.15a}$$

i.e.,

$$R(\sigma_\rho^f(C)) = \tilde{C}\tilde{R}(\rho^\ell)C \tag{A2.15b}$$

As emphasized in [17], this is an anti-isomorphism between $SO^\ell(3)$ and $SO^f(3)$, with

$$R(\sigma_{\rho\rho'}^f(C)) = R(\sigma_{\rho'}^f(C))R(\sigma_\rho^f(C)) \ . \tag{A2.16}$$

An anti-isomorphism can also be defined in the reverse direction by defining, for all ρ^f in $SO^f(3)$, a lab-fixed rotation $\sigma_\rho^\ell(C) \in SO^\ell(3)$ such that

$$R(\sigma_\rho^\ell(C))C = C\tilde{R}(\rho^f) \tag{A2.17a}$$

$$R(\sigma_\rho^\ell(C)) = C\tilde{R}(\rho^f)\tilde{C} \tag{A2.17b}$$

Contrary to the assertion by Louck and Galbraith ([17], VIC), we consider that the anti-isomorphism A2.17 is not relevant for the interpretation of Hougen/Longuet-Higgins PI theory. This point is discussed further in Chapter 2, §2.3.

The Hamiltonian for an arbitrary rotor is given by [43]

$$2\hat{H} = I^{-1}_{ij}\,\hat{K}_i\hat{K}_j \qquad \text{A2.18}$$

where I is the effective inertia tensor. Under an internal transformation ρ^f, the Hamiltonian becomes

$$2\hat{H}' = (I')^{-1}_{ij}\,\hat{K}_i\hat{K}_j \qquad \text{A2.19a}$$

where the new inertia tensor I' is given by

$$I' = \tilde{R}(\rho^f)IR(\rho^f) \qquad \text{A2.19b}$$

and is not in general equal to I. For a <u>spherical</u> rotor, however, $I' = I$ for <u>all</u> ρ^f. Since the external symmetry $SO^\ell(3)$ is valid for all systems in isotropic space, the symmetry group of the spherical rotor is therefore the product group $SO^\ell(3) * SO^f(3)$ (recall the constraint A2.6e). The symmetry $SO^\ell(3) * SO^f(3)$ explains the apparently 'accidental' degeneracies in spherical (and symmetric) top energy levels ([170], §IV).

We now obtain the generators of the group $SO^\ell(3) * SO^f(3)$. Consider first the group $SO^\ell(3)$ of external rotations. Let ψ be a function of the matrix C, transforming under operations of $SO^\ell(3)$ as follows (cf. Appendix 1):

$$\rho^\ell : \psi \to \hat{\Omega}_{\rho^\ell}\psi\ , \qquad \text{A2.20a}$$

$$[\hat{\Omega}_{\rho^\ell}\psi](C) = \psi(\tilde{R}(\rho^\ell)C)\ . \qquad \text{A2.20b}$$

Writing ρ^ℓ in the form $\rho^\ell = \rho^\ell(\theta,\hat{\underset{\sim}{n}})$, the generators of $SO^\ell(3)$ are defined as the 3 components of the vector $\hat{\underset{\sim}{J}}$, where

$$[\hat{\Omega}_{\rho^\ell}\psi] \equiv \exp(-\theta\hat{\underset{\sim}{n}}\cdot\hat{\underset{\sim}{J}})\psi\ . \qquad \text{AZ2.21}$$

Taking an infinitesimal rotation of $\delta\theta$ about the axis $\hat{\underset{\sim}{\ell}}_{k'}$ and expanding A2.21 to first order in $\delta\theta$, we have

$$\hat{\Omega}_{\rho^\ell}\psi \sim (1 - i\delta\theta\,\hat{J}_{k'})\psi \qquad \text{A2.21}$$

with $\hat{J}_{k'} \equiv \hat{\underset{\sim}{\ell}}_{k'}\cdot\hat{\underset{\sim}{J}}$. However, from equation A1.7

$$R(\rho^{\ell}(\delta\theta,\hat{\underset{\sim}{\ell}}_{k'}))_{ij} \sim \delta_{ij} - \delta\theta\varepsilon_{ijk}\delta_{kk'} \qquad \text{A2.22}$$

so that

$$\rho^{\ell}(\delta\theta,\hat{\underset{\sim}{\ell}}_{k'}):C_{is} \rightarrow (\delta_{ij} - \delta\theta\varepsilon_{ijk}\delta_{kk'})C_{js} \qquad \text{A2.23}$$

and

$$[\hat{\Omega}_{\rho^{\ell}}\psi](C) = \psi(\tilde{R}(\rho^{\ell}(\delta\theta,\hat{\underset{\sim}{\ell}}_{k'}))C) \sim \psi(C) + \delta\theta\varepsilon_{ijk}\delta_{kk'}C_{js}\partial\psi/\partial C_{is} \quad . \qquad \text{A2.24}$$

Comparison of equations A2.21 and A2.24 shows that

$$\hat{\mathbf{J}}_i = -i\varepsilon_{ijk}C_{js}\ \partial/\partial C_{ks} \qquad \text{A2.25}$$

which is the desired result: the generators of the group $SO^{\ell}(3)$ of external rotations are the lab-fixed components $\hat{\mathbf{J}}_i$ of the total orbital angular momentum. An important relation is

$$[\hat{\mathbf{J}}_i, C_{js}] = i\varepsilon_{ijk}C_{ks} \qquad \text{A2.26}$$

independent of s. This is the so-called co-rotation condition [30], and shows that the definition of the Eckart frame ensures that the molecule-fixed vectors $\hat{\underset{\sim}{f}}_i$ behave as true vectors with respect to the lab frame.

A similar calculation yields

$$\hat{\mathbf{K}}_i = -i\varepsilon_{iji}\tilde{C}_{js}\ \partial/\partial\tilde{C}_{ks} , \qquad \text{A2.27}$$

i.e., as expected, the generators of the group $SO^{f}(3)$ of internal rotations are the molecule-fixed components $\hat{\mathbf{K}}_i$ of the angular momentum. The result corresponding to A2.26 is

$$[\hat{\mathbf{K}}_i, \tilde{C}_{js}] = i\varepsilon_{ijk}\tilde{C}_{ks} \qquad \text{A2.28}$$

independent of s, which shows that the $\hat{\underset{\sim}{\ell}}_i$ behave as true vectors with respect to molecule-fixed rotations. Equations A2.26 and A2.28 are the starting point for the construction of rotational wavefunctions as polynomials in the C_{ij}, and this task is taken up in the next section.

Up until now, we have dealt only with the group $SO^{\ell}(3) * SO^{f}(3)$ of <u>proper</u> transformations of rotational variables. It is however a simple matter to extend this group to $O^{\ell}(3) * O^{f}(3)$, which is the symmetry group of the spherical rotor including improper external and internal rotations. To do this, we include the inversion transformation

$$\mathcal{I} : C \to - C \qquad \text{A2.29}$$

which reverses the relative sense of the lab- and molecule-fixed frames. $\mathcal{I}$ should be regarded as the inversion operation, since it does not matter whether we choose to regard it as an element of $O^{\ell}(3)$ or $O^{f}(3)$: the natural anti-isomorphism between the two groups map $\mathcal{I}$ onto itself.

The parity of molecular wavefunctions is defined by their behaviour under ,
and we stress again that the group of transformations of rotational variables associated with the molecular point group G is a subgroup G^f of $O^f(3)$

$$O^{\ell}(3) * O^{f}(3) \supset O^{\ell}(3) * G^{f} \supset G^{f} \quad . \qquad \text{A2.30}$$

A2.3. Construction of Rotational Wavefunctions

Recalling the relations A2.26 and A2.28, we see that the direction-cosine matrix C is a double vector with respect to the commuting angular momenta $\hat{\underset{\sim}{J}}$ (external indices) and $\hat{\underset{\sim}{K}}$ (internal indices).

Let $\langle i|1m\rangle$ be the unitary transformation relating the cartesian (T_i) and spherical (T^1_m) components of a vector, or spherical tensor of rank one [165]

$$T^1_m = \sum_i T_i \langle i|1m\rangle \qquad m = 0, \pm 1 \qquad \text{A2.31a}$$

$$T_i = \sum_m T^1_m \langle 1m|i\rangle \qquad i = x,y,z \qquad \text{A2.31b}$$

where

$$\sum_i \langle 1m|i\rangle\langle i|1m'\rangle = \delta_{mm'} \qquad \text{A2.32a}$$

$$\sum_m \langle i|1m\rangle\langle 1m|i\rangle = \delta_{ii'} \quad . \qquad \text{A2.32b}$$

In the Condon-Shortley phase convention,

$$\langle i|1m\rangle = \begin{array}{c} \\ \langle x| \\ \langle y| \\ \langle z| \end{array} \begin{array}{c} \begin{array}{ccc} |11\rangle & |10\rangle & |1-1\rangle \end{array} \\ \begin{bmatrix} -1/\sqrt{2} & 0 & 1/\sqrt{2} \\ -i/\sqrt{2} & 0 & -i/\sqrt{2} \\ 0 & 1 & 0 \end{bmatrix} \end{array} \qquad \text{A2.33}$$

so that the unitarity condition is

$$\langle 1m|i\rangle = \langle i|1m\rangle^* = (-)^m \langle i|1-m\rangle \quad . \qquad \text{A2.34}$$

Consider now the quantity

$$D^{(11)}_{mm'}(C) \equiv \sum_{ii'} C_{ii'} \langle i|1m\rangle\langle i'|1m'\rangle \qquad \text{A2.35}$$

(A2.35 is not a similarity transformation of C, but treats internal and external indices on the same footing.) Since C is a double-vector, we obtain directly

$$\hat{J}_z D^{(11)}_{mm'} = m D^{(11)}_{mm'} \qquad \text{A2.36a}$$

$$\hat{K}_z D^{(11)}_{mm'} = m' D^{(11)}_{mm'} \qquad \text{A2.36b}$$

$$\hat{J}_\pm D^{(11)}_{mm'} = [2 - m(m \pm 1)]^{1/2} D^{(11)}_{m\pm1,m'} \qquad \text{A2.37a}$$

$$\hat{K}_\pm D^{(11)}_{mm'} = [2 - m'(m' \pm 1)]^{1/2} D^{(11)}_{mm'\pm1} \qquad \text{A2.37b}$$

$$(\hat{J}_i \hat{J}_i) D^{(11)}_{mm'} = (\hat{K}_i \hat{K}_i) D^{(11)}_{mm'} = 2 D^{(11)}_{mm'} \quad , \qquad \text{A2.38}$$

i.e., $D^{(11)}$ is a spherical double-tensor of rank one, with the matrix C as argument; in other words, a rotational double-tensor (A2.6; cf. [11]).

For any C ε SO(3), the corresponding Wigner rotation matrix of rank one can be obtained by a similarity transformation of C (convention: Brink and Satchler [165])

$$\mathscr{D}^1(C)_{mm'} = \sum_{ii'} \langle 1m|i\rangle C_{ii'} \langle i'|1m'\rangle \qquad \text{A2.39}$$

so that

$$D^{(11)}_{mm'}(C) = (-)^{m'} \mathscr{D}^1(C)^*_{m-m'} \qquad \text{A2.40}$$

Our strategy now is to couple the double-vectors $D^{(11)}$ in standard fashion to give double tensors of higher rank j, which are proportional to suitably normalized rotational wavefunctions $|j;m,k\rangle$. The identity A2.74 then provides the well known relation between rotational wavefunctions and Wigner rotation matrices.

We note that the angular momentum operators A2.1 can be written in the form

$$\hat{J}_i = \sum_{\substack{\nu\nu' \\ \mu\mu'}} D^{(11)}_{\nu\mu} \langle 1;\nu,\mu|\hat{J}_i|1;\nu',\mu'\rangle\ \partial/\partial D^{(11)}_{\nu'\mu'} \qquad \text{A2.41}$$

$$\hat{J}_i = \sum_{\substack{\nu\nu' \\ \mu\mu'}} D^{(11)}_{\nu\mu} \langle 1;\nu,\mu|\hat{K}_i|1;\nu',\mu'\rangle\ \partial/\partial D^{(11)}_{\nu'\mu'}\ , \qquad \text{A2.42}$$

which has been generalized to arbitrary (integral or half-integral) tensor rank by Gulshani in his work on oscillator-like coherent states of the asymmetric top [16].

Let us then consider the coupled function

$$D^{(jj)}_{mm'} = \sum D^{(11)}_{m_1m_1'} D^{(11)}_{m_2m_2'} \langle 1m_1 1m_2|jm\rangle\langle 1m_1' 1m_2'|jm'\rangle \qquad \text{A2.43a}$$

$$= \sum C_{i_1i_1'} C_{i_2i_2'} \langle i_1|1m_1\rangle\langle i_2|1m_2\rangle\langle 1m_1 1m_2|jm\rangle$$

$$\langle i_1'|m_1'\rangle\langle i_2'|1m_2'\rangle\langle 1m_1' 1m_2'|jm'\rangle \qquad \text{A2.43b}$$

$$= \sum C_{i_1i_1'} C_{i_2i_2'} \langle i_1 i_2|1j;m\rangle\langle i_1' i_2'|1j;m'\rangle \quad , \qquad \text{A2.43c}$$

which is, by virtue of the definition A2.43a, a spherical double-tensor of rank j. In A2.43c, we have introduced the Cartesian-Spherical (CS) transformation coefficients defined by Stone [171]. Use of the relations given in Appendix V of [165] shows that the double-tensor $D^{(jj)}$ is related to the Wigner rotation matrix by

$$D^{(jj)}_{mm'}(C) = (-)^{1+1-j+m'} \mathcal{D}^{j}(C)_{m-m'} \qquad \text{A2.44}$$

so that for j = 2 we have a relation analogous to A2.40. These considerations can be generalized immediately to the case of n coupled double-tensors. Using the CS coefficients, we write

$$D^{(jj)}_{mm'}(C) = \sum C_{i_1i_1'} \cdots C_{i_ni_n'} \langle i_1\cdots i_n|\eta j;m\rangle\langle i_1' \cdots i_n'|\eta j;m'\rangle \qquad \text{A2.45}$$

where η denotes the particular coupling scheme $(j_1 = 1, j_{12},\ldots,j_{12\,\ldots\,n-1})$, and

$$\langle i_1 i_2 \cdots i_n | \eta j; m\rangle \equiv \sum \langle i_1 | 1m_1\rangle\langle i_2 | 1m_2\rangle \cdots \langle i_n | 1m_n\rangle$$

$$\langle 1m_1 1m_2 | j_{12} m_{12}\rangle \cdots \langle j_{12\cdots n-1} m_{12\cdots n-1} 1m_n | jm\rangle . \qquad A2.46$$

We shall consider only the unique 'fully-stretched' coupled tensors, for which $\eta \equiv (j_1 = 1,\ j_2 = 2, \ldots, j_{12\cdots n-1} = n-1)$, $j_n \equiv j = n$, and ([172], equation 61)

$$D^{(11)}_{mm'}(C) = (-)^{m'} \mathcal{D}^{j}(C)^{*}_{m-m'}, \quad \det C = +1. \qquad A2.47$$

The double-tensor $D^{(jj)}$ is therefore a homogeneous jth degree polynomial in the elements of the direction-cosine matrix C, given by A2.45 (the expression A2.45 is necessarily of rather formal significance; the creation/annihilation operator approach of Gulshani [16] is probably of greater practical utility).

To obtain normalized rotational wavefunctions $|j;m,k\rangle$, we use the standard orthogonality relation ([165], p. 147)

$$\int \mathcal{D}^{j_1}(C)^{*}_{m_1 m_1'}\, \mathcal{D}^{j_2}(C)_{m_2 m_2'}\, d\underline{\omega}(C) = \left[\frac{8\pi^2}{2j+1}\right] \delta_{j_1 j_2} \delta_{m_1 m_2} \delta_{m_1' m_2'} \qquad A2.48$$

so that (to within an arbitrary j-dependent phase)

$$\langle C | j;m,k\rangle = \left[\frac{2j+1}{8\pi^2}\right]^{1/2} D^{(jj)}_{mk}(C) \qquad A2.49$$

are rotational double tensors satisfying A2.6 and the orthonormality relations

$$\langle j;m,k | j';m',k'\rangle = \delta_{jj'} \delta_{mm'} \delta_{kk'} \quad . \qquad A2.50$$

The integral A2.48 is over the parameter domain of the matrix group SO(3), and is usually given in terms of the Euler angles. In fact, the work leading up to A2.49 appears to entail an unavoidable restriction to proper matrices $C \in SO(3)$. However, the crux of the method is that the expression A2.45 provides an unambiguous specification of the behaviour of rotational wavefunctions under the inversion $\mathcal{J}$. For, it is easily seen that, according to A2.45

$$D^{(jj)}(-C) = (-)^{j} D^{(jj)}(C) \qquad A2.51$$

and the double-tensors A2.49 are <u>parity eigenstates</u> with parity $\pi = (-)^{j}$, and may be denoted

$$\langle C|j^\pi;m,k\rangle, \qquad \pi = (-)^j \tag{A2.52}$$

Moreover, the function

$$(\det C)\, D^{(jj)}(C) \tag{A2.53}$$

obviously has the <u>opposite parity</u> $\pi = (-)^{j+1}$. It follows that for arbitrary j there is no difficulty in defining rotational wavefunctions $|j^\pi;m,k\rangle$ having <u>either</u> parity, + or -

$$\langle -C;j^\pi;m,k\rangle = (-)^{\nu_\pi}\langle C|j^\pi;m,k\rangle \tag{A2.54a}$$

where

$$\nu_\pi \equiv \begin{cases} 0 & \pi = g(+) \\ 1 & \pi = u(-) \end{cases} \quad . \tag{A2.54b}$$

This result is crucial, and shows that there are no essential difficulties associated with the use of discontinuous rotational wavefunctions defined on the matrix group O(3) (cf. [92,99,100]). The orthonormality relation becomes

$$\langle j^\pi;m,k|j'^{\pi'};m',k'\rangle = \delta_{jj'}\delta_{mm'}\delta_{kk'}\delta_{\pi\pi'} \quad . \tag{A2.50'}$$

Our final task is to consider the transformation properties of the $|j^\pi;m,k\rangle$ under the group $O^\ell(3) * O^f(3)$. We take first an external rotation $\rho^\ell \equiv (\rho^\ell, \rho_0^f)$, and set

$$R(\rho^\ell) \equiv (\det R(\rho^\ell))R'(\rho^\ell) \tag{A2.55}$$

so that $R'(\rho^\ell)$ is the proper part of $R(\rho^\ell)$. From A2.11 we see that

$$\rho^\ell:\langle C|j^\pi;m,k\rangle \to \langle(\det R(\rho^\ell))\tilde{R}'(\rho^\ell)C|j^\pi;m,k\rangle = (\det R(\rho^\ell))^{\nu_\pi}\langle\tilde{R}'(\rho^\ell)C|j^\pi;m,k\rangle \tag{A2.56}$$

To determine $\langle\tilde{R}'(\rho^\ell)C|j^\pi;m,k\rangle$, we have from A2.45

$$\begin{aligned} D^{(jj)}_{mk}(\tilde{R}'C) &= \sum (\tilde{R}'C)_{i_1i_1'}\cdots(\tilde{R}'C)_{i_ni_n'}\langle i_1\cdots i_n|\eta j;m\rangle\langle i_1'\cdots i_n'|\eta j;k\rangle \\ &= \sum \tilde{R}'_{i_1k_1}\cdots\tilde{R}'_{i_nk_n}\delta_{k_1k_1'}\cdots\delta_{k_nk_n'}C_{k_1'i_1'}\cdots C_{k_n'i_n'} \\ &\qquad \langle i_1\cdots i_n|\eta j;m\rangle\langle i_1'\cdots i_n'|\eta j;k\rangle \quad . \end{aligned} \tag{A2.57}$$

Inserting a 'complete set' of CS coefficients

$$\delta_{k_1 k_1'} \cdots \delta_{k_n k_n'} = \sum_{\eta' j' m'} \langle k_1 \cdots k_n | \eta' j' ; m' \rangle \langle \eta' j' ; m' | k_1' \cdots k_n' \rangle \qquad \text{A2.58}$$

and noting that cross terms in A2.57 must vanish ($\eta = \eta'$, $j = j'$), we find

$$\begin{aligned} D^{(jj)}_{mk}(\tilde{R}'C) &= \sum_{m'} D^{(jj)}_{m-m'}(\tilde{R}')(-)^{m'} D^{(jj)}_{m'k}(C) \\ &= \sum_{m'} \mathcal{D}^{j}(\tilde{R}')^{*}_{mm'} D^{(jj)}_{m'k}(C) \\ &= \sum_{m'} D^{(jj)}_{m'k}(C) \mathcal{D}^{j}(R')_{m'm} \quad , \end{aligned} \qquad \text{A2.59}$$

i.e.,

$$\rho^{\ell} : |j^{\pi};m,k\rangle \rightarrow \sum_{m'} |j^{\pi};m',k\rangle (\det R(\rho^{\ell}))^{\nu_{\pi}} \mathcal{D}^{j}(R'(\rho^{\ell}))_{m'm} \quad . \qquad \text{A2.60}$$

As expected, the lab-fixed rotation ρ^{ℓ} affects only the external index m (cf. [92] equation 3.57).

A similar calculation gives, for all $\rho^{f} \equiv (\rho^{\ell}_{0}, \rho^{f})$,

$$\rho^{f} : |j^{\pi};m,k\rangle \rightarrow \sum_{k'} |j^{\pi};m,k'\rangle (\det R(\rho^{f}))^{\nu_{\pi}} \mathcal{D}^{j}(R'(\rho^{f}))_{k'k} \qquad \text{A2.61}$$

which shows that the molecule-fixed rotation ρ^{f} affects only the internal index k (cf. [92], equation 3.33).

Combining the two equations A2.60 and 61, we obtain the main result of this appendix: for all (ρ^{ℓ}, ρ^{f}) in $O^{\ell}(3) * O^{f}(3)$,

$$\begin{aligned} (\rho^{\ell}, \rho^{f}) : |j^{\pi};m,k\rangle \rightarrow \sum |j^{\pi};m',k'\rangle (\det R(\rho^{\ell}))^{\nu_{\pi}} (\det R(\rho^{f}))^{\nu_{\pi}} \cdot \\ \mathcal{D}^{j}(R'(\rho^{\ell}))_{m'm} \mathcal{D}^{j}(R'(\rho^{f}))_{k'k} \quad . \end{aligned} \qquad \text{A2.62}$$

This equation forms the basis for our discussion of the effect of molecular symmetry operations on rotational wavefunctions in Chapter 2.

Appendix 3

The Class Structure and Irreducible Representations of Semi-direct Product Groups

In Chapter 3, we have shown how semi-direct product structure may arise in NRM groups. Apart from its formal significance, the presence of semi-direct product structure is of great practical utililty. Thus, as discussed in detail by Altmann [125] (cf. also [139, 147, 151]), both the class structure and the irreducible representations of a semi-direct product group can be calculated using systematic procedures that exploit the (known) properties of the smaller component groups to the maximum extent. Such techniques represent an optimal use of purely group-theoretical information in the NRM problem and, as shown by the applications described in Chapter 4, can provide an effective approach to the classification of large NRM symmetry groups.

It is our purpose in this appendix to state Altmann's results on semi-direct products ([125], Chapter 19) in a form we have found convenient for the calculations of Chapter 4; for all proofs etc. we refer to [125]. Some changes have been made in Altmann's notation in order to bring it into line with that of our Chapter 3. The semi-direct product nature of the point symmetry groups is discussed in Appendix 4.

A3.1 Vocabulary

1) Invariant Subgroups

Consider the group **G**, with elements $g_1, g_2, \ldots$, and the group **K**, with elements s,t.

For any $t \in \mathbf{K}$, $g \in \mathbf{G}$, the <u>conjugate element</u> g^t is

$$g^t \equiv tgt^{-1} \equiv tg\bar{t} \qquad \text{A3.1}$$

which is assumed defined. The <u>conjugate group</u> $\mathbf{G}^t$ is the set

$$\mathbf{G}^t \equiv \{g^t \mid g \in \mathbf{G}\} \quad . \qquad \text{A3.2}$$

The group G is said to be invariant under $t \in K$ if

$$G^t = G \ . \tag{A3.3}$$

The group G is invariant under K, written $G \lessdot K$, if the conjugate group G^t equals G for all elements of K, i.e.,

$$G \lessdot K \ \text{if} \ G^t = G \quad \forall\, t \in K \ . \tag{A3.4}$$

If G is a subgroup of H, $G \subset H$, and G is invariant under H, then G is an invariant subgroup of H, written

$$G \lhd H \ . \tag{A3.5}$$

2) Vector and Projective Representations

A vector representation $\hat{G}$ of a group G is a set of matrices $\{\hat{G}(g) \mid g \in G\}$ obeying the multiplication rule

$$\hat{G}(g_1)\hat{G}(g_2) = \hat{G}(g_1 g_2) \tag{A3.6}$$

for all g_1, g_2 in G. This is the usual definition of a group representation. Irreducible representations (IRs) of the group G are denoted ${}^iG$.

A projective or ray representation G of a group G is a set of matrices $\{\check{G}(g) \mid g \in G\}$ obeying the multiplication rule

$$\check{G}(g_1)\check{G}(g_2) = [g_1, g_2]\check{G}(g_1 g_2) \ , \tag{A3.7}$$

where the number $[g_1, g_2]$ is the projective factor, and the $|G|^2$ constants $[g_i, g_j] \equiv [i,j]$ $(g_i, g_j \in G)$ form the factor system. The projective factors $[i,j]$ must satisfy the associativity condition

$$[i,j][ij,k] = [i,jk][j,k] \tag{A3.8}$$

for all g_i, g_j and g_k in G.

For a finite group, there is a finite number of (so-called gauge) equivalence classes of factor systems: without loss of generality, we may take the projective factors to be complex numbers of modulus unity

$$|[i,j]| = 1 \ . \tag{A3.9}$$

Irreducibililty of projective representations is defined exactly as for ordinary vector representations [115].

3) The Semi-direct Product

Consider two groups G and **K**, where $G \lessdot K$. The complex of $|G||K|$ (distinct) elements $[t|g]$ with composition law

$$[t_1|g_1][t_2|g_2] = [t_1t_2|\ g_1^{\bar{t}_2}\ g_2] \qquad \text{A3.10a}$$

i.e., Identity: $[e_G|e_K]$ A3.10b

and Inverse: $[t|g]^{-1} = [\bar{t}|\bar{g}^{t}]$ A3.10c

is a group H, the semi-direct product of G and K, denoted

$$H = G \textcircled{\wedge} K . \qquad \text{A3.11}$$

Note that G is an invariant subgroup of H,

$$G \lhd H . \qquad \text{A3.12}$$

If $K \lessdot G$ as well, then H is the usual direct product group

$$H = G \otimes K = K \otimes G \quad . \qquad \text{A3.13}$$

4) Little Co-groups and Orbits

Consider the semi-direct product $G \textcircled{\wedge} K$. Let $\hat{G}$ be a vector representation of G. The conjugate representation $\hat{G}^t$ is defined by

$$\hat{G}^t(g) \equiv \hat{G}(g^{\bar{t}}) \qquad \text{A3.14}$$

for all g in G, t in K. Let ${}_i\hat{G}$ be an IR of G. The little co-group ${}_i\mathcal{K}$ of the IR ${}_i\hat{G}$ is the set of all $t \in K$ such that ${}_i\hat{G}^t$ is equivalent to ${}_i\hat{G}$

$${}_i\mathcal{K} \equiv \{t \in K|\,{}_i\hat{G}^t \sim {}_i\hat{G}\} \quad , \qquad \text{A3.15}$$

and is a (proper or improper) subgroup of K

$$ {}_i\mathcal{K} \subset \mathbf{K} \qquad \text{A3.16} $$

The <u>little group</u> ${}_iK$ is the semi-direct product

$$ {}_iK \equiv G \circledS \mathcal{K}_i \subset G \circledS K \ . \qquad \text{A3.17} $$

Note that

$$ {}_i\hat{G}^{[t|g]} \sim {}_i\hat{G} \quad \forall\ [t|g] \in {}_iK \ . \qquad \text{A3.18} $$

The <u>orbit</u> of the representation ${}_i\hat{G}$, denoted ${}_i\mathbf{S}$, is the set of all conjugate representations ${}_i\hat{G}^s$ which are mutually <u>inequivalent</u>. To generate the orbit, we form the coset decomposition of $\mathbf{K}$ with respect to the little co-group ${}_i\mathcal{K}$

$$ \mathbf{K} = \sum_{\nu=1}^{|\nu|} s_\nu\, {}_i\mathcal{K} \qquad |\nu| = |\mathbf{K}|/|{}_i\mathcal{K}| \qquad \text{A3.19a} $$

and have

$$ {}_iS = \{{}_i\hat{G}^{s_\nu}\} \ . \qquad \text{A3.19b} $$

A given orbit is determined by any one of its <u>prongs</u> ${}_i\hat{G}^{s_\nu}$.

5) Induced Representations

Given: the groups G, H, with G ⊂ H; the set of functions $\{\phi_r;\ r = 1\cdots|\hat{G}|\}$ spanning the representation $\hat{G}$; the coset decomposition of **H** with respect to **G**

$$ \mathbf{H} = \sum_{\sigma=1}^{|\sigma|} s_\sigma G \qquad |\sigma| = |\mathbf{H}|/|\mathbf{G}| \qquad \text{A3.20} $$

The set of $|\hat{G}|\cdot|\mathbf{H}|/|\mathbf{G}|$ functions $\{S_\sigma\phi_r;\sigma = \cdots|\sigma|;\ r = 1\cdots|\hat{G}|\}$, which are assumed to be linearly independent (this is the case of <u>regular</u> induction: cf. [173]), are said to span the <u>induced representations</u> of $\hat{G}$ onto H, denoted $\hat{G} \uparrow H$. In detail, the induced representation is constructed as follows: forming the <u>ground representation</u> of H, denoted ${}^{\gamma}\hat{H}$

$$ {}^{\gamma}\hat{H}(h)_{\sigma\sigma'} = \begin{cases} 1 & \text{if } hs_{\sigma'}G = s_{\sigma}G \\ 0 & \text{otherwise} \end{cases} \qquad \text{A3.21a} $$

$$ |{}^{\gamma}\hat{H}| = |\sigma| = |\mathbf{H}|/|\mathbf{G}| \quad , \qquad \text{A3.21b} $$

and defining

$$ g_{\sigma'}(h) \equiv \sum_{\sigma=1}^{|\sigma|} \bar{s}_{\sigma}\, {}^{\gamma}\hat{H}(h)_{\sigma\sigma'} hs_{\sigma'} \qquad \text{A3.22} $$

the induced representation matrices are

$$ \hat{G} \uparrow H(h)_{\sigma r,\sigma' r'} = {}^{\gamma}\hat{H}(h)_{\sigma\sigma'} \hat{G}[g_{\sigma'}(h)]_{rr'} \qquad \text{A3.23a} $$

where $r,r' = 1 \cdots |\hat{G}|$; $\sigma,\sigma' = 1 \cdots |\sigma|$, with characters

$$ \chi(h|\hat{G} \uparrow H) = \sum_{\sigma} {}^{\gamma}\hat{H}(h)_{\sigma\sigma}\, \chi(\bar{s}_{\sigma} hs_{\sigma}|\hat{G}) \quad . \qquad \text{A3.23b} $$

A3.2 The Class Structure of $G \circledS \mathbf{K}$

The class structure of a semi-direct product group $G \circledS \mathbf{K}$ is determined very easily using the algorithms shown in Figures A3.1 and A3.2, which summarize the contents of [125], §19.10. The simplified procedure in Figure A3.2 is appropriate when the invariant subgroup G is Abelian; further simplifications obtain when either element s or element g is the identity, or when **K** is cyclic.

The following centralizer groups are defined:

$$ P \equiv Z(g|\mathbf{G}) \equiv \{g' \in \mathbf{G} | g^{g'} = g\} \qquad \text{A3.24} $$

$$ M \equiv Z(s|\mathbf{G}) \equiv \{g' \in \mathbf{G} | (g')^{s} = g'\} \qquad \text{A3.25} $$

$$ N \equiv Z(g|M) = M \cap P \qquad \text{A3.26} $$

$$ A \equiv Z(\mathbf{G}|\mathbf{K}) = \{t \in \mathbf{K} | g^{t} = g \text{ for all } g \in \mathbf{G}\} \qquad \text{A3.27} $$

$$ B \equiv Z(s|\mathbf{K}) = \{t \in \mathbf{K} | s^{t} = s\} \qquad \text{A3.28} $$

together with the coset expansions

$$G = \sum_{x} xM \qquad \text{A3.29}$$

$$M = \sum_{y} yN \qquad \text{A3.30}$$

$$\mathbf{K} = \sum_{\delta} A\delta B = \sum_{\xi} \xi B \qquad \text{A3.31}$$

$$= \sum_{r\delta} \eta_r^{\delta}\delta B \qquad \text{A3.32}$$

$$\{\eta_r^{\delta}\delta\} \equiv \{\xi\} \qquad \text{A3.33}$$

where $$A \equiv \sum_{r} \eta_r^{\delta} L^{\delta} \qquad \text{A3.34}$$

with $$L^{\delta} \equiv A \cap B^{\delta} \quad . \qquad \text{A3.35}$$

Also,

$$C[s|g] \equiv \text{Class of } [s|g] \text{ in } G \circledwedge \mathbf{K} \qquad \text{A3.36}$$

$$C(s) \equiv \text{Class of } s \text{ in } \mathbf{K} \quad . \qquad \text{A3.37}$$

The equations A3.32 and A3.33 relate the single and double coset expansions of **K** shown in A3.31. In all the cases we have treated so far, (Chapter 4) the single and double coset expansions have been trivially related.

The object of this seemingly complicated procedure is to avoid the immense and unenlightening labour of a direct application of the conjugation relation

$$[s_i|g_i]^{[s_j|g_j]} = [s_j s_i \bar{s}_j | (g_j^{\bar{s}_i} g_i)^{s_j} \bar{g}_j^{s_j}] \qquad \text{A3.38}$$

for all pairs of elements $[s_i|g_i]$, $[s_j|g_j]$ of $G \circledwedge \mathbf{K}$.

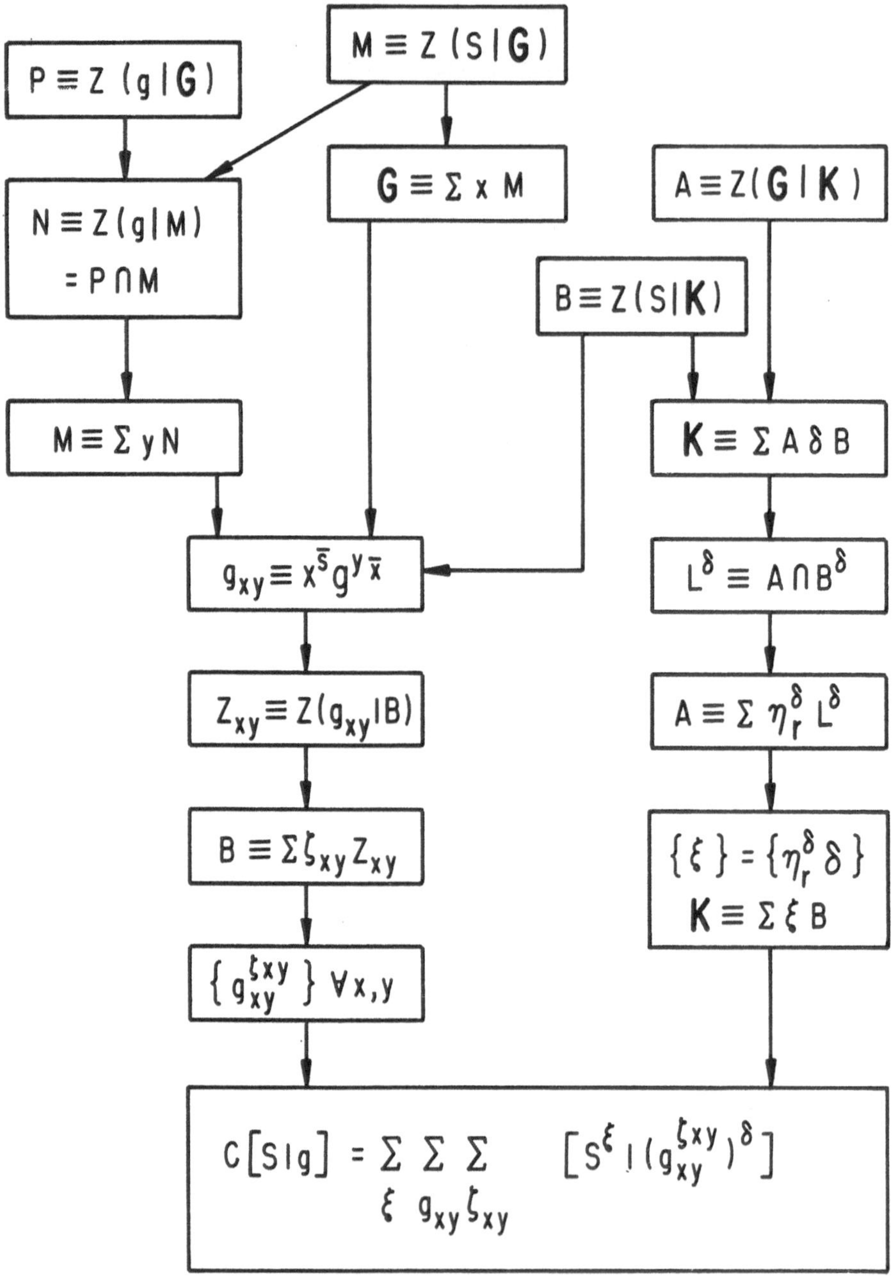

Figure A3.1 Algorithm for determining the class structure of the semi-direct product G Ⓢ K.

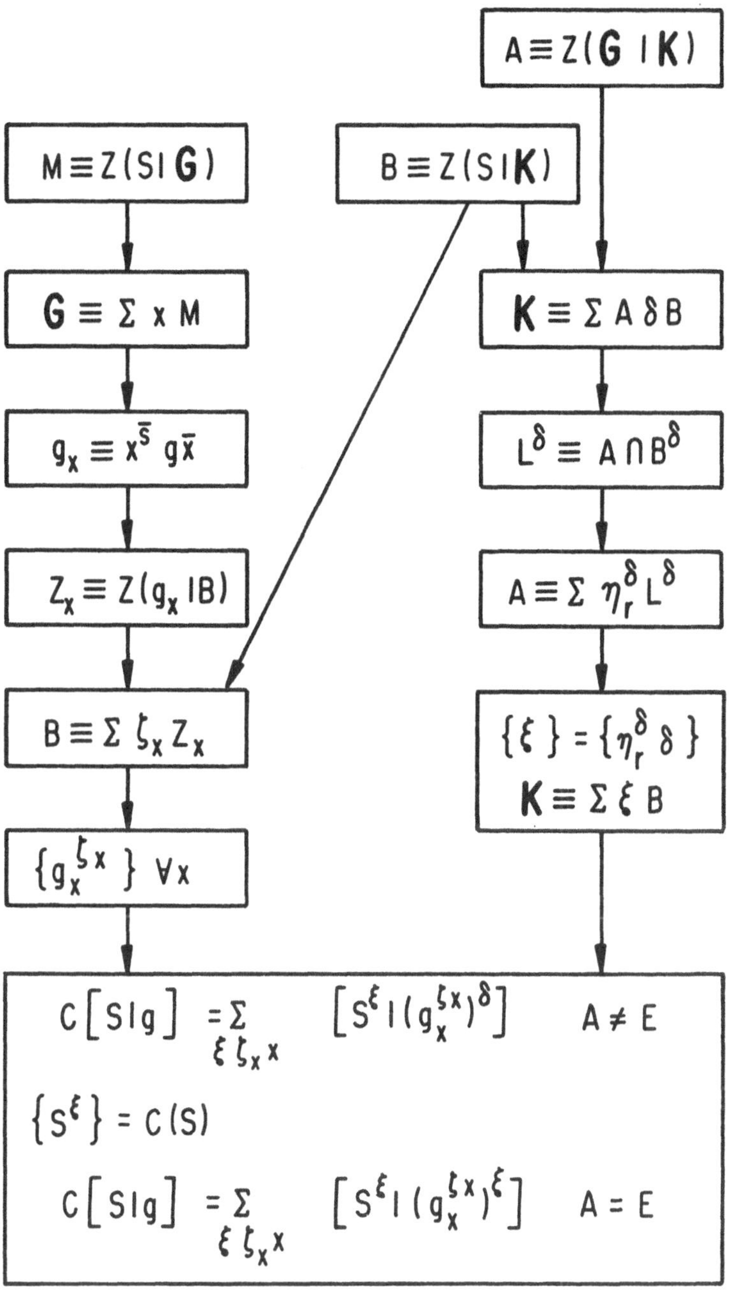

Figure A3.2 Algorithm for determining the class structure of **G** ⊗ **K**, where **G** is abelian.

A3.3 The Irreducible Representations of G Ⓐ K

The irreducible representations of the semi-direct product group H = G Ⓐ K are obtained as follows:

1) Classify the IRs of the group G, $\{{}_i\hat{G}\}$, into orbits (A3.19).

2) Choose one IR ${}_i\hat{G}$ from each orbit, and find the little co-group ${}_i\mathcal{K}$ (A3.15). The little group is ${}_iK \equiv G Ⓐ {}_i\mathcal{K}$ (A3.17).

3) Find a representation of ${}_i\mathcal{K}$, ${}_i\check{\mathcal{K}}$, such that

$$ {}_i\check{\mathcal{K}}(t)^{-1}\,{}_i\hat{G}(g)\,{}_i\check{\mathcal{K}}(t) = {}_i\hat{G}^t(g) \qquad \forall\, t \in {}_i\mathcal{K} \,. \tag{A3.39} $$

This is the connecting representation [147]. Note that ${}_i\check{\mathcal{K}}$ is in general a projective representation:

$$ {}_i\check{\mathcal{K}}(s)\,{}_i\check{\mathcal{K}}(t) = [s,t]\,{}_i\check{\mathcal{K}}(st) \quad . \tag{A3.40} $$

However, in all the cases we have treated so far (Chapter 4), the connecting representations have been vector representations of ${}_i\mathcal{K}$.

4) Choose an irreducible (projective) representation of ${}_i\mathcal{K}$, denoted ${}^u_i\check{\mathcal{K}}$, such that

$$ {}^u_k\check{\mathcal{K}}(s)\ {}^u_i\check{\mathcal{K}}(t) = [s,t]'\ {}^u_i\check{\mathcal{K}}(st) \tag{A3.41a} $$

where

$$ [s,t]' \equiv [s,t]^{-1} \tag{A3.41b} $$

i.e., the factor system is the inverse of that associated with the connecting representation. In all the cases we have considered, these have therefore been the usual vector IRs of the little co-group.

5) Form the vector IR of the little group

$$ {}^u_i\hat{K}[s|g] \equiv {}^u_i\check{\mathcal{K}}(s) \otimes {}_i\check{\mathcal{K}}(s)\,{}_i\hat{G}(g) \qquad s \in {}_i\mathcal{K},\ g \in G \tag{A3.42a} $$

with characters

$$ \chi([s|g] \mid {}^u_i\hat{K}) = \chi(s|{}^u_i\check{\mathcal{K}})\ \mathrm{Tr}\{{}_i\check{\mathcal{K}}(s)\ {}_i\hat{G}(g)\} \quad . \tag{A3.42b} $$

This vector IR is a permitted small representation (see [125]).

6) Induce the permitted small representation ${}^{u}_{i}\hat{K}$ found in 5) from ${}_{i}K$ to $\mathbf{H}$. That is,

a) Form the coset decomposition

$$\mathbf{H} = \sum_{\tau=1}^{|\tau|} t_{\tau}\,{}_{i}K \qquad |\tau| = |\mathbf{K}|/|{}_{i}\mathcal{K}| \quad . \qquad \text{A3.43}$$

b) Form the ground representation of $\mathbf{H}$

$$\Upsilon\hat{H}[s|g]_{tt'} = \begin{cases} 1 & \bar{t}st' \in {}_{i}\mathcal{K} \\ 0 & \bar{t}st' \notin {}_{i}\mathcal{K} \end{cases} \quad . \qquad \text{A3.44}$$

c) Form the induced representation ${}^{u}_{i}\hat{K}{\uparrow}H$, <u>which is an IR of **H**</u>, ${}^{u}_{i}\hat{K} \uparrow H \equiv {}^{u}_{i}\hat{H}$

$${}^{u}_{i}\hat{H}[s|g]_{[tt']} \equiv {}^{\Upsilon}\hat{H}[s|g]_{tt'}\{{}^{u}_{i}\check{\mathcal{K}}(\bar{t}st') \otimes {}_{i}\check{\mathcal{K}}(\bar{t}st')\ {}_{i}\hat{G}^{t}(g)\} \qquad \text{A3.45a}$$

with characters

$$\chi([s|g]|\ {}^{u}_{i}\hat{H}) = \sum_{g:\ s^{\bar{t}} \in {}_{i}\mathcal{K} \text{ or } s \in {}_{i}\mathcal{K}^{t}} \chi(s|{}^{n}_{i}\mathcal{K}^{t})\ \mathrm{Tr}\{{}_{i}\mathcal{K}^{t}(s) \otimes {}_{i}\hat{G}^{t}(g)\} \ . \qquad \text{A3.45b}$$

Each IR of H therefore carries the label of an orbit (i) and an IR of ${}_{i}\mathcal{K}(u)$. A central result is that, upon taking all distinct orbits i in turn, together with all the corresponding permitted small representations, we obtain <u>all</u> the vector IRs of H <u>once and once only</u>.

This procedure is very easy to implement in practice,and provides a labeling scheme for the IRs of NRM symmetry groups that contains a great deal of useful information, as can be seen from the rigid/nonrigid correlation diagrams constructed in Chapter 4.

Appendix 4

The Point Groups as Semi-direct Products

In this appendix the semi-direct product structure of the point groups is described, and it is shown how the general procedures summarized in Appendix 3 lead to a systematic determination of point group irreducible representations. McIntosh [146] has emphasized that the resulting nomenclature for the IRs is perhaps more rational than the familiar notation established by Placzek (cf. [6,85]); the point symmetry groups have also been treated from this viewpoint by Altmann [125,174], Bradley and Cracknell [175], and Fritzer [176,177] (see also [186,187]). The semi-direct decomposition of the point groups is helpful in the construction of rigid/nonrigid correlation diagrams, as discussed in Chapter 4.

A4.1 Cyclic and Dihedral Groups

Some generic relations between the cyclic, dihedral and related point groups are shown in Figure A4.1, where the notation

$$C_n = \{\hat{E} \equiv \hat{C}^n_{nz}, \hat{C}_{nz}, \ldots, \hat{C}^{n-1}_{nz}\} \qquad \text{A4.1}$$

$$C_{2'} = \{\hat{E}, \hat{C}_{2y}\} \qquad \text{A4.2}$$

$$C_s = \{\hat{E}, \hat{\sigma}_{yz}\} \qquad \text{A4.3}$$

$$C_i = \{\hat{E}, \hat{i}\} \qquad \text{A4.4}$$

$$C_h = \{\hat{E}, \sigma_{xy}\} \qquad \text{A4.5}$$

$$C_{s'} = \{\hat{E}, \hat{\sigma}'\} \qquad \text{A4.6}$$

is used, coordinate axes being defined as in Figure A4.2 (cf. [125], §19.6).

Figure A4.1 Some generic relations between the Cyclic and Dihedral point groups (ascent in symmetry)

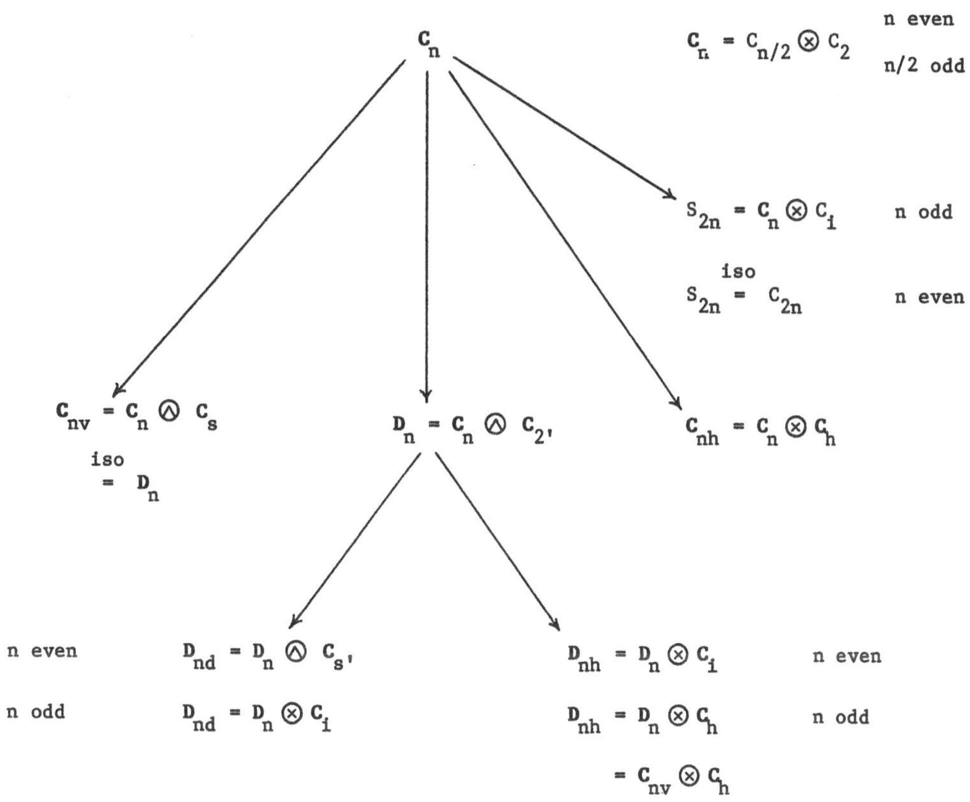

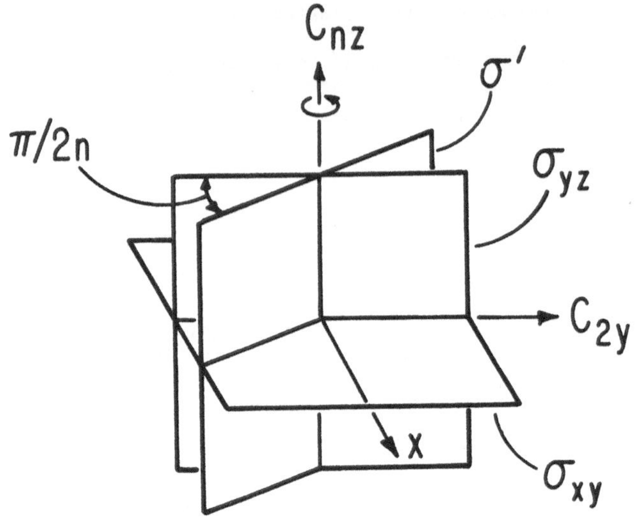

Figure A4.2 Coordinate axes and symmetry elements for cyclic and dihedral point groups.

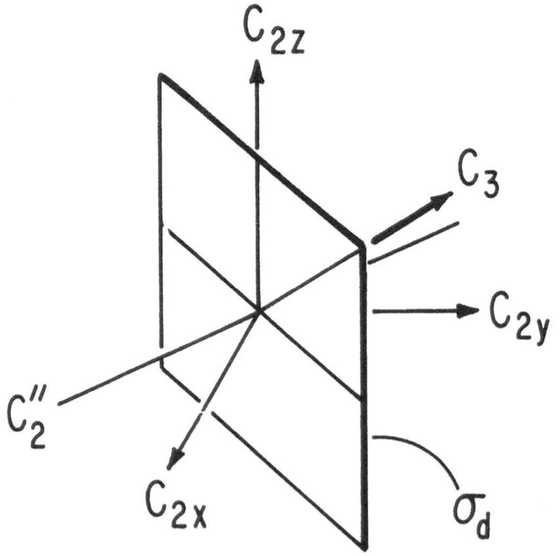

Figure A4.3 Coordinate axes and symmetry elements for cubic point groups.

a) Cyclic Groups

The character table of the Abelian cyclic group C_n has the form

C_n	$\hat{E}$	$\hat{C}_{nz}$	$\hat{C}^2_{nz}$	...	$\hat{C}^{n-1}_{nz}$
(0)	1	1	1	...	1
(1)	1	ε	ε^2	...	ε^{n-1}
(m)	1	ε^m	ε^{2m}	...	$\varepsilon^{(n-1)m}$
(n-1)	1	ε^{n-1}	$\varepsilon^{2(n-1)}$	...	$\varepsilon^{(n-1)^2}$

where the IRs are denoted by a number (m) (m taken <u>mod</u> n), $\varepsilon = \exp(i2\pi/n)$, and the character

$$\chi(\hat{C}^m_{nz}|(m')) = \varepsilon^{mm'}. \tag{A4.7}$$

b) Dihedral Groups.

The dihedral group D_n is the semidirect product

$$D_n = C_n \circledS C_2, \,. \tag{A4.8}$$

The orbits and class structure, hence the form of the character table, depend upon whether n is even or odd. In both cases, however, we have the conjugation relations (cf. Appendix 3):

Elements: $$(\hat{C}^m_{nz})^{\hat{C}_{2y}} = \hat{C}^{-m}_{nz} \equiv \hat{C}^{n-m}_{nz} \quad , \tag{A4.9a}$$

IRs: $$(m)^{\hat{C}_{2y}} = (-m) \equiv (n-m) \quad . \tag{A4.9b}$$

i) n odd

Class Structure: $\{\hat{E}\}$

$\{\hat{C}^{i}_{nz}, \hat{C}^{n-i}_{nz}\}$ $i = 1,\ldots,d \equiv (n-1)/2$

$\{\hat{C}_{2y}\hat{C}^{i}_{nz} \mid i = 1,\ldots,n\}$

Orbits of C_n: $\{0\}$ ${}_0\mathcal{K} = C_{2'}$

$\{(i), (n-i)\}$ $i = 1,\ldots,d;$ ${}_i\mathcal{K} = \{\hat{E}\}$

The character table of D_n for n odd is

D_n	1	2	2		2	n
	E	$\hat{C}_{nz}$	$\hat{C}^2_{nz}$		$\hat{C}^d_{nz}$	$\hat{C}_{2y}$
$A_1 \equiv (0)A_1$	1	1	1	...	1	1
$A_2 \equiv (0)A_2$	1	1	1	...	1	-1
$E_1 \equiv (1)A$	2	$2c(\omega)$	$2c(2\omega)$	...	$2c(d\omega)$	0
$E_2 \equiv (2)A$	2	$2c(2\omega)$	$2c(4\omega)$	...	$2c(2d\omega)$	0
⋮						
$E_d \equiv (d)A$	2	$2c(d\omega)$	$2c(2d\omega)$	...	$2c(d^2\omega)$	0

where $\omega \equiv 2\pi/n$ and $c(\omega) \equiv \cos(\omega)$. The conventional (A,E) notation for the irreducible representations is given as well as the more systematic semi-direct product notation obtained by writing the IR of the invariant subgroup C_n specifying a given orbit together with an IR of the corresponding little co-group (see Appendix 3). The two IRs of the group $C_{2'}$ are denoted A_1 and A_2, respectively.

ii) n even

Class Structure

$\{\hat{E}\}$

$\{\hat{C}^{i}_{nz},\ \hat{C}^{n-i}_{nz}\}\quad i = 1,\ldots,(n-2)/2$

$\{\hat{C}^{n/2}_{nz}\}$

$\{\hat{C}_{2y}\hat{C}^{i}_{nz} \mid i = 0,2,\ldots,n-2\}$

$\{\hat{C}_{2y}\hat{C}^{i}_{nz} \mid i = 1,3,\ldots,n-1\}$

Orbits of $\mathbf{C}_n$

$\{(0)\}$ $\qquad {}_0\mathcal{K} = \mathbf{C}_{2'}$

$\{(i),\ (n-i)\}\quad i = 1,\ldots,(n-2)/2$ $\qquad {}_i\mathcal{K} = \{\hat{E}\}$

$\{(n/2)\}$ $\qquad {}_{n/2}\mathcal{K} = \mathbf{C}_{2'}$

The character table of $\mathbf{D}_n$ for n even is $(f \equiv (n-2)/2)$:

$\mathbf{D}_n$	1	2	2	...	1	n/2	n/2
	$\hat{E}$	$\hat{C}_{nz}$	$\hat{C}^{2}_{nz}$	...	$\hat{C}^{n/2}_{nz}$	$\hat{C}_{2y}$	$\hat{C}_{2y}\hat{C}_{nz}$
$A_1 \equiv (0)A_1$	1	1	1	...	1	1	1
$A_2 \equiv (0)A_2$	1	1	1	...	1	-1	-1
$B_1 \equiv (n/2)A_1$	1	-1	1	...	$(-)^{n/2}$	1	-1
$B_2 \equiv (n/2)A_2$	1	-1	1	...	$(-)^{n/2}$	-1	1
$E_1 \equiv (1)A$	2	$2c(\omega)$	$2c(2\omega)$	...	$2c(n\omega/2)$	0	0
$E_f \equiv (f)A$	1	$2c(f\omega)$	$2c(2f\omega)$	...	$2c(fn\omega/2)$	0	0

c) Cyclic Groups with reflection

$C_{nh} = C_n \otimes C_h$	Direct product for all n.
$C_{nv} = C_n \circledS C_s$	Semi-direct product, isomorphic with corresponding group D_n.

d) Dihedral Groups with reflection

n odd:	$D_{nh} = D_n \otimes C_h$	Direct product
	$D_{nd} = D_n \otimes C_i$	Direct product
n even:	$D_{nh} = D_n \otimes C_i$	Direct product
	$D_{nd} = D_n \circledS C_{s'}$	Semi-direct product

The class structure of the semi-direct product $D_{nd} = D_n \circledS C_{s'}$ (n even) is:

	Number of Classes	Number of Elements
$\{\hat{E}\}$, $\{\hat{C}^m_{nz}, \hat{C}^{n-m}_{nz}\}$ $m = 1,\ldots,f$, $\{\hat{C}^{n/2}_{nz}\}$	$(n/2) + 1$	n
$\{\hat{C}_{2y}\hat{C}^m_{nz} \mid m = 0,1,\ldots,n-1\}$	1	n
$\{\hat{\sigma}'\hat{C}^m_{nz} \mid m = 0,1,\ldots,n-1\}$	1	n
$\{\hat{\sigma}'\hat{C}_{2y}\hat{C}^m_{nz}, \hat{\sigma}'\hat{C}_{2y}\hat{C}^{-m-1}_{nz}\}$ $m = 0,1,\ldots,f$	$n/2$	n
Total	$n + 3$	4n

The orbits of D_n under $C_{s'}$ and associated IRs of D_{nd} are:

Orbit	Little Co-group	IRs of $\mathbf{D}_{nd}$ [6]
$A_1 \equiv (0)A_1$	${}_{A_1}\mathcal{K} = C_{s'}$	A_1, B_1
$A_2 \equiv (0)A_2$	${}_{A_2}\mathcal{K} = C_{s'}$	A_2, B_2
$E_1 \equiv (1)A$	${}_{E_1}\mathcal{K} = C_{s'}$	E_1, E_{n-1}
$E_f \equiv (f)A$	${}_{E_f}\mathcal{K} = C_{s'}$	E_f, E_{n-f}
$B_1 \equiv (n/2)A_1$, $B_2 \equiv (n/2)A_2$	${}_{B_1}\mathcal{K} = \{\hat{E}\}$	$E_{f/2}$

A4.2 Cubic Groups

With axes defined as in Figure A4.3, the cubic groups have the following structure:

$$\mathbf{T} = \mathbf{D}_2 \textcircled{\scriptsize$\wedge$} \mathbf{C}_3 \quad \text{A4.10}$$

$$\mathbf{T}_d = \mathbf{D}_2 \textcircled{\scriptsize$\wedge$} \mathbf{C}_{3v} \quad \text{A4.11a}$$

$$= \mathbf{D}_2 \textcircled{\scriptsize$\wedge$} (\mathbf{C}_3 \textcircled{\scriptsize$\wedge$} \mathbf{C}_s(\sigma_d)) \quad \text{A4.11b}$$

$$\mathbf{T}_h = \mathbf{T} \otimes \mathbf{C}_i \quad \text{A4.12}$$

$$\mathbf{O} = \mathbf{D}_2 \textcircled{\scriptsize$\wedge$} (\mathbf{C}_3 \textcircled{\scriptsize$\wedge$} \mathbf{C}_{2''}) \quad \text{A4.13a}$$

$$= \mathbf{D}_2 \textcircled{\scriptsize$\wedge$} \mathbf{D}_{3'} \quad \text{A4.13b}$$

$$\mathbf{O}_h = \mathbf{O} \textcircled{\scriptsize$\wedge$} \mathbf{C}_i \quad \text{A4.14}$$

where conjugation relations are

$$\hat{C}_{2x}^{\hat{C}_3^+} = \hat{C}_{2y} \quad \text{A4.15a}$$

$$\hat{C}_{2x}^{\hat{\sigma}_d} = \hat{C}_{2y} \quad \text{A4.15b}$$

and so on. Consider the example of the tetrahedral group. The class structure of $\mathbf{T}_d$ is:

E	$\hat{E}$
$3C_2$	$\hat{C}_{2x}, \hat{C}_{2y}, \hat{C}_{2z}$
$8C_3$	$\hat{C}_3, \hat{C}_3\hat{C}_{2x}, \hat{C}_3\hat{C}_{2y}, \hat{C}_3\hat{C}_{2z}$ $\hat{C}_3^2, \hat{C}_3^2\hat{C}_{2x}, \hat{C}_3^2\hat{C}_{2y}, \hat{C}_3^2\hat{C}_{2z}$
$6\sigma_d$	$\hat{\sigma}_d, \hat{\sigma}_d\hat{C}_{2z}$ $\hat{\sigma}_d\hat{C}_3, \hat{\sigma}_d\hat{C}_3\hat{C}_{2x}$ $\hat{\sigma}_d\hat{C}_3^2, \hat{\sigma}_d\hat{C}_3^2\hat{C}_{2y}$
$6S_4$	$\hat{\sigma}_d\hat{C}_{2x}, \hat{\sigma}_d\hat{C}_{2y}$ $\hat{\sigma}_d\hat{C}_3 C_{2y}, \hat{\sigma}_d\hat{C}_3\hat{C}_{2z}$ $\hat{\sigma}_d\hat{C}_3^2 C_{2z}, \hat{\sigma}_d\hat{C}_3^2\hat{C}_{2x}$

and orbits of $\mathbf{D}_2$ under $\mathbf{C}_{3v}$ are

$$\{(0)A_1\} \qquad {}_{0A_1}\mathcal{K} = \mathbf{C}_{3v}$$

$$\{(0)A_2,\ (1)A_1,\ (1)A_2\} \qquad {}_{0A_2}\mathcal{K} = \mathbf{C}_s$$

In semi-direct product form the character table of $\mathbf{T}_d$ is:

$\mathbf{T}_d$	1	3	8	6	6
C_{3v}	$\hat{E}$	$\hat{E}$	$\hat{C}_3$	$\hat{\sigma}_d$	$\hat{\sigma}_d$
D_2	$\hat{E}$	$\hat{C}_{2x}$	$\hat{E}$	$\hat{E}$	$\hat{C}_{2x}$
$A_1 \equiv (0)A_1\cdot(0)A'$	1	1	1	1	1
$A_2 \equiv (0)A_1\cdot(0)A''$	1	1	1	-1	-1
$E \equiv (0)A_1\cdot(1)A$	2	2	-1	0	0
$T_2 \equiv (0)A_2\cdot A'$	3	-1	0	1	-1
$T_1 \equiv (0)A_2\cdot A''$	3	-1	0	-1	1

We have the immediate correlation between the IRs of T_d and those of its subgroups D_2 and C_{3v}:

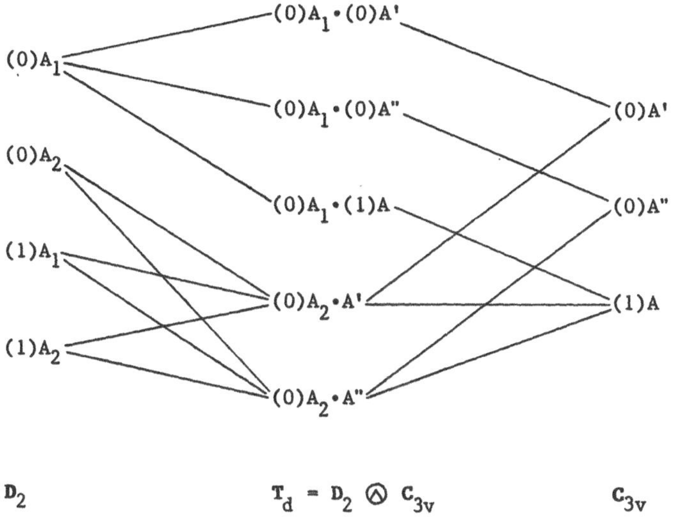

A4.3 Icosahedral Groups

The icosahedral group, being simple, has no nontrivial invariant subgroup and hence no semi-direct product structure.

Glossary of Symbols and Abbreviations

Mathematical notation

$\exists$	there exists
$\forall$	for all
$\{x\|P\}$	the set of elements x having the property P
$x \in X$	x is an element of the set X
$A \subset B$	A is a subset (subgroup) of the set (group) B
$A \lhd B$	A is an invariant subgroup of the group B
$a_i b_i \equiv \sum_i a_i b_i$	Summation convention for repeated indices i,j,k, ...
ε_{ijk}	Antisymmetric unit tensor: $\varepsilon_{ijk} = +1$ if (ijk) = (xyz) or any cyclic permutation thereof; $\varepsilon_{ijk} = -1$ if (ijk) = (yxz) or any cyclic permutation thereof; $\varepsilon_{ijk} = 0$ otherwise.
$(a \wedge b)_i = \varepsilon_{ijk} a_j b_j$	vector product

Abbreviations:

NRM	Nonrigid molecule
SRMM	Semirigid molecular model
IR	Irreducible representation
PI	Permutation - inversion
CNPI	Complete nuclear permutation - inversion
PPT	Primitive period transformation

References

1. Longuet-Higgins, H.C., 1963, Molec. Phys., 6, 445.
2. Buck, B., Biedenharn, L.C., & Cusson, R.Y., 1979, Nucl. Phys., A317, 205.
3. Sutcliffe, B.T., 1979, Lectures presented at NATO ASI 'Quantum Dynamics of Molecules', Cambridge, England.
4. Makushkin, Yu. S., & Ulenikov, O.N., 1977, J. Mol. Spec., 68, 1.
5. Sørensen, G.O., 1979, Topics in Current Chemistry, 82, 99.
6. Wilson, E.B., Decius, J.C., & Cross, P.C., 1955, Molecular Vibrations (McGraw-Hill).
7. Louck, J.D., 1976, J. Mol. Spec., 61, 107.
8. Meyer, R., & Günthard, Hs. H., 1968, J. Chem. Phys., 49, 1510.
9. Eckart, C., 1935, Phys. Rev., 47, 552.
10. Sayvetz, A., 1939, J. Chem. Phys., 7, 383.
11. Judd, B.R., 1975, Angular Momentum Theory for Diatomic Molecules (Academic).
12. Husson, N., 1975, Annls Phys., 9, 271.
13. Meyer, F.O., & Redding, R.W., 1978, J. Mol. Spec., 70, 410.
14. Hilico, J-C, Berger, H., & Loete, M., 1976, Can. J. Phys., 54, 1702.
15. Pascaud, E., & Poussigue, G., 1978, Can. J. Phys., 56, 1577.
16. Gulshani, P., 1979, Can. J. Phys., 57, 998.
17. Louck, J.D., & Galbraith, H.W., 1976, Rev. Mod. Phys., 48, 69.
18. Nielsen, H.H., 1951, Rev. Mod. Phys., 23, 90.
19. Howard, B.J., & Moss, R.E., 1970, Mol. Phys., 19, 433.
20. Born, M., & Oppenheimer, R., 1927, Ann. Physik, 84, 457.

20a. Kiselev, A.A., 1978, Can. J. Phys., 56, 615.

21. Özkam, I., & Goodman, L., 1979, Chem. Rev., 79, 275.
22. Jørgensen, F., & Pedersen, T., 1974, Mol. Phys., 27, 33.
23. Pedersen, T., 1978, Unpublished Report, University of Copenhagen.
24. Podolsky, B., 1928, Phys. Rev., 32, 812.
25. Wilson, E.B., & Howard, J.B., 1936, J. Chem. Phys., 4, 260.
26. Watson, J.K.G., 1968, Mol. Phys., 15, 479.
27. Watson, J.K.G., 1970, Mol. Phys., 19, 465.
28. Whittaker, E.T., 1929, Analytical Dynamics (Cambridge).
29. Malhiot, R.J., & Ferigle, S.M., 1954, J. Chem. Phys., 22, 717.
30. Herold, H., & Ruder, H., 1974, J. Phys., G5, 341.
31. Woolley, R.G., 1976, Adv. Phys., 25, 27.
32. Woolley, R.G., 1978, J. Am. Chem. Soc., 100, 1073.
33. Woolley, R.G., 1979, Israel J. Chem., 18, No. 4.
34. Dalton, B.J., & Nicholson, P.D., 1975, Int. J. Quant. Chem., 9, 525.
35. Liehr, A.D., 1963, Prog. Inorg. Chem., 5, 385.
36. Goldstein, H., 1950, Classical Mechanics (Addison-Wesley).
37. Gulshani, P., & Rowe, D.J., 1976, Can. J. Phys., 54, 970.
38. Peric, M., 1977, Mol. Phys., 34, 1675.
39. Natanson, G.A., & Adamov, M.N., 1974, Vest. Leningr. Univ., No. 10, 24; Jørgensen, F., 1978, Int. J. Quant. Chem., 24, 55.
40. Quade, C.R., 1976, J. Chem. Phys., 64, 2783.
41. Hoy, A.R., Mills, I.M., & Strey, G., 1972, Mol. Phys., 24, 1265.
42. Barakat, R., 1979, Mol. Phys., 38, 1655.
43. Kroto, H.W., 1975, Molecular Rotation Spectra (Wiley).
44. Essen, H., 1978, Am. J. Phys., 46, 983.
45. Jørgensen, F., 1978, Thesis, University of Copenhagen.
46. Rowe, D.J., & Rosensteel, G., 1979, J. Math. Phys., 20, 465.
47. Casimir, H.B.G., 1931, Rotation of a Rigid Body in Quantum Mechanics (Walters).
48. Bauder, A., Meyer, R., & Günthard, Hs. H., 1974, Mol. Phys., 28, 1305.
49. Ezra, G.S., 1979, Mol. Phys., 38, 863.
50. Hougen, J.T., Bunker, P.R., & Johns, J.W.C., 1970, J. Mol. Spec., 34, 136.
51. Hougen, J.T., 1964, Can. J. Phys., 42, 1920.
52. Hougen, J.T., 1965, Can. J. Phys., 43, 935.
53. Bunker, P.R., 1967, J. Chem. Phys., 47, 718.
54. Papousek, D., & Spirko, V., 1976, Topics Current Chem., 68, 59.
55. Russeger, P., & Brickmann, J., 1975, J. Chem. Phys., 62, 1086.

56. Gilles, J.M.F., & Philippot, J., 1972, Int. J. Quant. Chem., 6 225.
57. Kreglewski, M., 1978, J. Mol. Spec., 72, 1.
58. Hougen, J.T., & Redding, R.W., 1971, J. Mol. Spec., 37, 366.
59. Sarka, K., 1971, J. Mol. Spec., 29, 66.
60. Bauder, A., & Gunthard, Hs. H., 1976, J. Mol. Spec., 60, 290.
61. Berry, R.S., 1979, Lectures presented at NATO ASI 'Quantum Dynamics of Molecules', Cambridge, England.
62. Abraham, F.F., 1974, Homogeneous Nucleation Theory (Academic).
63. Kendrick, J., & Hillier, I.H., 1977, Mol. Phys., 33, 635.
64. Martin, R.L., & Davidson, E.R., 1978, Mol. Phys., 35, 1713.
64a. Gerber, W.H., & Schumacher, E., 1978, J. Chem. Phys., 69, 1692.
65. Berry, R.S., 1960, J. Chem. Phys., 32, 933.
66. Whitesides, G.M., & Mitchell, H.L., 1969, J. Am. Chem. Soc., 91, 5384.
67. Berry, R.S., Personal Communication.
68. Dalton, B.J., 1966, Mol. Phys., 11, 265.
69. Dalton, B.J., 1971, J. Chem. Phys., 54, 4745.
70. Brocas, J., & Fastenakel, D., 1975, Mol. Phys., 30, 193; Dalton, B.J., Brocas, J., & Fastenakel, D., 1976, Mol. Phys., 31, 1887; Fastenakel, D., & Brocas, J., 1980, Mol. Phys., 40, 361; Brocas, J., Fastenakel, D., and Buschen, J., 1980, Mol. Phys., 41, 1163.
71. Trindle, C., Data, S.N., & Bouman, T.D., 1977, Int. J. Quant. Chem., 11, 627.
72. Curtiss, C.F., Hirschfelder, J.O., & Adler, F.T., 1950, J. Chem. Phys., 18, 1638.
73. Eckart, C., 1934, Phys. Rev., 46, 384.
74. Van Vleck, J.H., 1935, Phys. Rev., 47, 487.
75. Bohr, A., 1976, Rev. Mod. Phys., 48, 365.
76. Fabre de la Ripelle, M., & Navarro, J., 1979, Ann. Phys., 123, 185.
77. Kramer, P., & Moshinsky, M., 1968, in Group Theory and its Applications, Vol. I, edited by E.M. Loebl (Academic).
78. Smirnov, Yu. F., & Shitikova, K.V., 1977, Sov. J. Part. Nucl., 8, 344.
79. Bosanac, S., & Murrell, J.N., 1973, Mol. Phys., 26, 349.
80. Wallace, R., 1979, Chem. Phys., 37, 93.
81. Kellmann, M.E., & Berry, R.S., 1976, Chem. Phys. Lett., 42, 327.
82. Amar, F., Kellmann, M.E., & Berry, R.S., 1979, J. Chem. Phys., 70, 1973; Amar, F., Kellmann, M.E., & Berry, R.S., 1980, J. Chem. Phys., 73, 2387; Ezra, G.S., & Berry, R.S., 1981, J. Chem. Phys., to be published.
83. Harter, W.G., Patterson, C.W., & da Paixao, F.J., 1978, Rev. Mod. Phys., 50, 37.
84. Wigner, E.P., 1930, Nachr. Ges. Wiss. Göttingen, page 133; Brester, C.J., 1923, Dissertation (Utrecht).
85. Herzberg, G., 1945., Infrared and Raman Spectra of Polyatomic Molecules (Van Nostrand).
86. Wilson, E.B., 1934, J. Chem. Phys., 2, 432.
87. Wilson, E.B., 1935, J. Chem. Phys., 3, 276.
88. Wilson, E.B., 1935, J. Chem. Phys., 3, 818.
89. Hougen, J.T., 1962, J. Chem. Phys., 37, 1433.
90. Hougen, J.T., 1963, J. Chem. Phys., 39, 358.
91. Hougen, J.T., 1971, J. Chem. Phys., 55, 1122.
92. Hougen, J.T., 1975, MTP International Review of Science, Vol. 3 (Physical Chemistry Series 2), page 75.
93. Mills, I.M., 1964, Mol. Phys., 7, 549.
94. Bunker, P.R., & Papousek, D., 1969, J. Mol. Spec., 32, 419.
95. Oka, T., 1973, J. Mol. Spec., 48, 503.
96. Bunker, P.R., 1975, Vibrational Spectroscopy and Structure, Vol. 3, edited by J.R. Durig (Marcel Dekker).
97. Bunker, P.R., 1979, Molecular Symmetry and Spectroscopy (Academic).
98. Moret-Bailly, J., 1965, Cahier Phys., 178, 253.
99. Moret-Bailly, J., 1974, J. Mol. Spec., 50, 483.
100. Hougen, J.T., 1974, J. Mol. Spec., 50, 485.
101. Jahn, J.A., 1938, Proc. R. Soc. A, 168, 469.
102. Michelot, F., Bobin, B., & Moret-Bailley, J., 1979, J. Mol. Spec., 76, 374.
103. Cantrell, C.D., 1976, in Physics of Quantum Electronics, Vol. 4 (Addison-

Wesley).
104. Cantrell, C.D., & Galbraith, H.W., 1975, J. Mol. Spec., 58, 158.
105. Berger, H., 1977, J. de Phys., 38, 1371.
106. Fano, U., & Chang, E.S., 1972, Phys. Rev., A6, 173.
107. Harter, W.G., & Patterson, C.W., 1976, A Unitary Calculus for Electronic Orbitals, Lecture Notes in Physics 49 (Springer Verlag).
108. Lathouwers, L.L., 1978, Phys. Rev. A18, 2150.
109. Herzberg, G., 1950, Spectra of Diatomic Molecules (Van Nostrand).
110. Townes, C.H., & Schawlow, A.L., 1955, Microwave Spectroscopy (McGraw-Hill).
111. Pfeifer, P., 1979, Preprint ETH-Zurich.
112. Gilles, J.M.F., & Philippot, J., 1978, Int. J. Quant. Chem., 14, 299.
113. Flurry, R.L., & Siddall, T.H., 1978, Mol. Phys., 36, 1309.
114. Pack, R.T., & Hirschfelder, J.O., 1968, J. Chem. Phys., 49, 4009.
115. Wigner, E.P., 1959, Group Theory (Academic).
116. Watson, J.K.G., 1967, J. Chem. Phys., 46, 1935.
117. Levy-Leblond, J.M., 1971, in Group Theory & its Applications, Vol. 2, edited by E.M. Loebl (Academic).
118. Watson, J.K.G., 1965, Can. J. Phys., 43, 1996.
119. Herzberg, G., 1966, Electronic Spectra and Electronic Structure (Van Nostrand).
120. McWeeny, R., 1963, Symmetry (Pergamon).
121. Butler, P.R., 1975, Phil. Trans. R. Soc., 277, 545.
122. Bunker, P.R., 1973, J. Mol. Spec., 48, 181.
123. Altmann, S.L., 1967, Proc. R. Soc. A, 298, 184.
124. Altmann, S.L., 1971, Mol. Phys., 21, 587.
125. Altmann, S.L., 1977, Induced Representations in Crystals and Molecules (Academic).
126. Natanson, G., 1976, Opt. Spectrosc., 41, 18.
127. Dellepiane, G., Gussoni, M., & Hougen, J.T., 1973, J. Mol. Spec., 47, 515.
128. Watson, J.K.G., & Merer, A.J., 1973, J. Mol. Spec., 47, 499.
129. Dyke, T.R., Howard, B.J., & Klemperer, W.A., 1972, J. Chem. Phys., 56, 2442.
130. Dyke, T.R., 1977, J. Chem. Phys., 66, 492.
131. Quack, M., 1977, Mol. Phys., 34, 477.
132. Metropoulos, A., & Chiu, Y-N., 1978, J. Chem. Phys., 68, 5607.
133. Hougen, J.T., 1965, Pure Appl. Chem., 11, 481.
134. Frei, H., Meyer, R., Bauder, A., & Günthard, Hs. H., 1976, Mol. Phys., 32, 43.
135. Frei, H., Groner, P., Bauder, A., & Günthard, Hs. H., 1978, Mol. Phys., 36, 1469.
136. Stone, A.J., 1964, J. Chem. Phys., 41, 1568.
137. Watson, J.K.G., 1971, Mol. Phys., 21, 577.
138. Woodman, C.M., 1970, Mol. Phys., 19, 753.
139. Serre, J., 1968, Int. J. Quant. Chem., 2S, 107.
140. Serre, J., 1974, Adv. Quantum Chem., 8, 1.
141. Wilson, E.B., Lin, C.C., & Lide, D.R., 1955, J. Chem. Phys., 23, 136.
142. Natanson, G.A., & Adamov, M.N., 1974, Vestn. Leningr. Gos. Univ. No. 4, 22.
143. Fleming, J.W., & Banwell, C.N., 1969, J. Mol. Spec., 31, 378.
144. Turrel, G., 1970, J. Mol. Struct., 5, 245.
145. Nourse, J.G., & Mislow, K., 1975, J. Am. Chem. Soc., 97, 4571.
146. McIntosh, H., 1960, J. Mol. Spec., 5, 269.
147. McIntosh, H., 1963, J. Mol. Spec., 10, 51.
148. Dreizler, H., 1968, Fortchr. Chem. Forsch., 10, 59.
149. Natanson, G.A., & Adamov, M.N., 1970, Vestn. Leningr. Univ., No. 4, 50.
150. Günthard, Hs.H., Bauder, A., & Frei, H., 1979, Topics in Current Chemistry, 81, 7.
151. Balasubramanian, K., 1980, J. Chem. Phys., 72, 665.
152. Coxeter, H.S.M., & Moser, W.O.J., 1972, Generators and Relations for Discrete Groups (Springer).
153. Magnus, W., Karrass, A., & Solitar, D., 1976, Combinatorial Group Theory (Dover).
154. Natanson, G.A., 1979, Opt. Spectrosc., 47, 137.
155. Bunker, P.R., 1965, Mol. Phys., 9, 257.
156. Bunker, P.R., 1964, Mol. Phys., 8, 81.

157. Bunker, P.R., 1965, Mol. Phys., 9, 247.
158. Hamada, Y., Hirawaka, A.Y., Tamagake, K., & Tsuboi, M., 1970, J. Mol. Spec., 35, 420.
159. Flurry, R.L., 1975, J. Mol. Spec., 56, 88.
160. Flurry, R.L., & Abdulnur, S.F., 1976, J. Mol. Spec., 63, 33.
161. Littlewood, D.E., 1950, The Theory of Group Characters (Oxford, Clarendon).
162. Wallace, R., 1979, Chem. Phys., 37, 285.
163. Gust, D., & Mislow, K., 1973, J. Am. Chem. Soc., 95, 1353.
164. Bunker, P.R., 1979, Lecture Notes in Chemistry, 12, 38.
165. Brink, D.M., & Satchler, G.R., 1968, Angular Momentum (Clarendon).
166. Rose, M.E., 1957, Elementary Theory of Angular Momentum (Wiley).
167. Dirac, P.A.M., 1958, The Principles of Quantum Mechanics (Clarendon).
168. Van Vleck, J.H., 1951, Rev. Mod. Phys., 23, 213.
169. Howard, B.J., & Brown, J.M., 1976, Mol. Phys., 31, 1517.
170. McIntosh, H.V., 1971, in Group Theory and its Applications, Vol. 2, edited by E.M. Loebl (Academic).
171. Stone, A.J., 1975, Mol. Phys., 29, 1461.
172. Fox, K., & Ozier, I., 1970, J. Chem. Phys., 52, 5044.
173. Mead, C.A., 1974, Topics in Current Chemistry, 49, 1.
174. Altmann, S.L., 1963, Phil. Trans. R. Soc., A255, 216.
175. Bradley, C.J., & Cracknell, A.P., 1972, The Mathematical Theory of Symmetry in Solids (Clarendon).
176. Fritzer, H.P., 1976, Lecture Notes in Physics, 50, 348.
177. Fritzer,H.P., 1979, in Recent Advances in Group Theory and their Applications to Spectroscopy, edited by J.C. Donini (Plenum: NATO ASI series).
178. Metropoulos, A., & Chiu, Y-N., 1979, Chem. Phys., 42, 61.
179. Klein, D.J., 1975, in Group Theory and its Applications, Vol. 3, edited by E.M. Loebl (Academic).
180. Seligman, T.H., 1979, Lecture Notes in Chemistry, 12, 178.
181. Louck, J.D., 1979, Lecture Notes in Chemistry, 12, 57.
182. Dress, A., 1979, Lecture Notes in Chemistry, 12, 77.
183. Natanson, G.A., & Adamov, M.N., 1976, 'New Aspects in the Symmetry Theory of Polyatomic Molecules' (Translated from Fiz. Molek. (USSR), Los Alamos translation LA-TR-77-32); Natanson, G.A., 1978, Fiz. Molek. (USSR), No. 6, 3.
184. Ezra, G.S., 1981, Mol. Phys., 43, 773.
185. Frei, H., Bauder, A., & Gunthard, Hs.H., 1981, Mol. Phys., 43, 785.
186. Melvin, M., 1956, Rev. Mod. Phys., 28, 18.
187. Boyle, L.L., & Walker, J.R., 1977, Int. J. Quant. Chem., 12 S2, 157.
188. Odutola, J.A., Alvis, D.L., Curtis, C.W., & Dyke, T.R., 1981, Mol. Phys., 42, 267.
189. Ezra, G.S., 1981, Submitted for publication.
190. Weber, A., 1980, J. Chem. Phys., 73, 3952.
191. Renkes, G.D., 1981, Chem. Phys., 57, 261.